NORTH CAROLINA
STATE BOARD OF COMMUNITY COLLEGES
LIBRARIES
ASHEVILLE-BUNCOMBE TECHNICAL COMMUNITY COLLEGE

PAVEMENT MAINTENANCE AND REHABILITATION

DISCARDED

A symposium sponsored by
ASTM Committee D-4 on
Road and Paving Materials
Bal Harbour, FL, 7 Dec. 1983

DEC - 6 2024

ASTM SPECIAL TECHNICAL PUBLICATION 881
Bernard F. Kallas, The Asphalt
Institute, editor

ASTM Publication Code Number (PCN)
04-881000-08

1916 Race Street, Philadelphia, PA 19103

Library of Congress Cataloging-in-Publication Data

Main entry under title:

Pavement maintenance and rehabilitation.

(ASTM special technical publication; 881)
Includes bibliographies and index.
"ASTM publication code number (PCN) 04-881000-08."
1. Pavements—Maintenance and repair—Congresses.
I. Kallas, Bernard F.   II. ASTM Committee D-4 on
Road and Paving Materials.   III. Series.
TE250.P285   1985   625.7′6   85-7550
ISBN 0-8031-0424-3

Copyright © by AMERICAN SOCIETY FOR TESTING AND MATERIALS 1985
Library of Congress Catalog Card Number: 85-7550

NOTE

The Society is not responsible, as a body,
for the statements and opinions
advanced in this publication.

Printed in Baltimore, MD
July 1985

Foreword

The symposium on Pavement Maintenance and Rehabilitation was presented at Bal Harbour, FL, on 7 Dec. 1983. The symposium was sponsored by ASTM Committee D-4 on Road and Paving Materials. Bernard F. Kallas was chairman of the symposium and is editor of this publication.

Related ASTM Publications

Placement and Compaction of Asphalt Mixtures, STP 829 (1984), 04-829000-08

Properties of Flexible Pavement Materials, STP 807 (1983), 04-807000-08

Asphalt Pavement Construction: New Materials and Techniques, STP 724 (1981), 04-724000-08

Quality Assurance in Pavement Construction, STP 709 (1980), 04-709000-08

Recycling of Bituminous Pavements, STP 662 (1978), 04-662000-08

A Note of Appreciation to Reviewers

The quality of the papers that appear in this publication reflects not only the obvious efforts of the authors but also the unheralded, though essential, work of the reviewers. On behalf of ASTM we acknowledge with appreciation their dedication to high professional standards and their sacrifice of time and effort.

ASTM Committee on Publications

ASTM Editorial Staff

Susan L. Gebremedhin
Janet R. Schroeder
Kathleen A. Greene
William T. Benzing

Contents

Introduction	1
Correcting Flexible Pavement Deficiencies: The Ontario Way— GEORGE J. CHONG AND WILLIAM A. PHANG	3
Evaluation of Asphaltic Pavements for Overlay Design— ROBERT C. DEEN, HERBERT F. SOUTHGATE, AND GARY W. SHARPE	18
Pavement Friction Measurements and Vehicle Control Reparations for Nontangent Road Sections—GLENN G. BALMER, RICHARD A. ZIMMER, AND RICHARD D. TONDA	33
Heavy Duty Membrane for the Reduction of Reflective Cracking in Bituminous Concrete Overlays—NORMAN E. KNIGHT AND GARY L. HOFFMAN	51
The Use of an Asphalt Polymer/Glass Fiber Reinforcement System for Minimizing Reflection Cracks in Overlays and Reducing Excavation Before Overlaying—R. DAVID ROWLETT AND WILLIAM E. UFFNER	65
The Use of a Gyratory Testing Machine in the Evaluation of Cold-Recycled Asphalt Paving Mixtures—MANG TIA AND LEONARD E. WOOD	74
Discussion	90
Index	95

Overview

Maintenance and rehabilitation of pavements become increasingly important as road systems grow and mature, and fewer new roads and streets are built. As road and street systems grow, maintenance and rehabilitation also become more complex and difficult to manage and finance because new pavements often deteriorate at widely differing rates depending on traffic, environment, materials, design, construction, and other factors. Some of the factors cannot be or are difficult to control, or are imperfectly understood. Maintenance and rehabilitation of the present network of roads and streets in the United States will require more engineering attention and increasingly larger proportions of the budgets of public agencies responsible for roads and streets for some time into the future.

This symposium was organized to provide authoritative information and useful current research findings to engineers and researchers in important areas of pavement maintenance and rehabilitation using bituminous materials. The several papers describe procedures for detecting and repairing defects in pavements and materials for repairing defects in pavements. Also covered are rehabilitation methods and equipment for evaluating the safety and structural adequacy of pavements, and design of improvements to satisfy current and future service requirements.

Of interest to all engineers responsible for maintenance of road systems is Chong and Phang's concise description of the Ontario Ministry of Transportation and Communications "Pavement Maintenance Guidelines" including distress classification methodology and procedures leading to the most cost-effective treatment. They also stress the need for more research on materials and ways for improving the effectiveness of sealing pavement cracks.

Deen et al provide information of particular interest and value to design engineers and researchers in a summary of a method developed through extensive research in Kentucky to evaluate the structural adequacy of asphalt concrete pavements before designing overlays. The procedures utilize elastic theory and dynamic pavement deflection measurements and have been simplified to the extent that overlay thickness requirements can be determined from a design nomograph. The authors have also used their procedures with dynamic deflection data obtained directly on subgrades, dense-graded aggregates, pozzolanic bases, full-depth asphalt concrete, and on portland cement concrete pavements, but these applications are considered experimental and subject to further study.

Useful information needed for safety considerations is provided by Balmer et al who address the measurement of friction deterioration on curves, ramps, intersections and other transitional areas, and the various remedial procedures to improve frictional characteristics.

Two papers present information of interest to all agencies using or considering use of various materials to reduce reflection cracking in asphalt concrete overlays. Knight and Hoffman describe the performance of seven different types of heavy duty membranes placed over portland cement concrete pavement joints at a site in Pennsylvania before placing of an asphalt concrete overlay. After two freezing seasons, they observed differences in the performance of the different types of membranes and found that the membranes as a whole were retarding reflection cracking. However, the cost/benefits of the membranes are still undetermined from their experiments. Rowlett and Uffner describe the performance of an asphalt polymer/glass fiber reinforcement system in 150 field trials in 21 states and concluded the system was effective in minimizing reflection cracking in asphalt overlays. As with the Knight and Hoffman experiments in Pennsylvania, the long-term cost effectiveness asphalt polymer/glass fiber system remains to be determined. The asphalt polymer/glass fiber system was also included among the several membranes evaluated by Knight and Hoffman.

Of particular interest to engineers and agencies using or conducting research on asphalt cold-mix recycling for pavement rehabilitation are Tia and Wood's experiments with the gyratory testing machine for evaluating cold-recycled asphalt paving mixtures. They conclude that the gyratory stability index and the gyratory elasto-plastic index determined in the fixed roller mode can be used to detect unstable mixtures when the binder content is too high. Their conclusions are based on correlations of gyratory testing machine results with several other commonly used mechanical test properties but were not correlated with the field performance of cold recycled mixtures.

Much information is presented in the papers of the symposium that is useful to agencies facing unprecedented pavement maintenance and rehabilitation demands. The symposium papers also identify many areas of pavement maintenance and rehabilitation where research and development should be continued or increased. It is evident that there is need and opportunity for many innovations in cost-effective materials, equipment, design, construction, and management systems for the maintenance and rehabilitation of pavements.

Bernard F. Kallas
The Asphalt Institute, College Park, MD 20740; symposium chairman and editor.

George J. Chong[1] and William A. Phang[1]

Correcting Flexible Pavement Deficiencies: The Ontario Way

REFERENCE: Chong, G. J. and Phang, W. A., "**Correcting Flexible Pavement Deficiencies: The Ontario Way,**" *Pavement Maintenance and Rehabilitation, ASTM STP 881*, B. F. Kallas, Ed., American Society for Testing and Materials, Philadelphia, 1985, pp. 3-17.

ABSTRACT: Ontario emphasizes preservation of their highway system by carrying out maintenance activities aimed at prolonging the life of highway pavements. To make the maintenance task easier and more effective, a set of guidelines was developed that helps maintenance staff make the best decision, one that is most cost-effective and one that is made consistently by different maintenance staff. For the most part, the guidelines deal with short-term measures-deficiencies that have to be fixed right now, that is, corrective maintenance. But they can also be used to identify situations where preventive maintenance will affect the service life of the pavement. The guidelines contain the following elements: (1) identification process, (2) treatment selection, and (3) performance standard.

This paper describes our current practice of identifying and classifying a typical deficiency, selection of the most cost-effective treatment, the specifications for equipment and materials needed to carry out the treatment, and the proper work methods. As a typical example, this paper will describe flexible pavement crackings and cracks that are routed and sealed as the prescribed treatment.

KEY WORDS: pavements, maintenance, highways, correcting flexible pavement deficiencies

Ontario emphasizes preservation of the King's Highway system by carrying out both corrective and preventive maintenance activities aimed at prolonging the life of highway pavements. A step was taken towards the achievement of this objective when the "Pavement Maintenance Guidelines" [1] were developed in 1980. The Guidelines incorporate two principal features:

1. A method of pavement deficiency identification combined with a consistent method of determining which of the available corrective actions is esti-

---

[1]Research engineer and head of pavement research, respectively, Ontario Ministry of Transportation and Communications, 1201 Wilson Ave., Downsville, Ontario, Canada M3M 1J8.

mated to be most cost-effective. "Corrective maintenance" is carried out to maintain the character and integrity of the pavement so as to ensure the safe operation of the pavement system.

2. Complementing these deficiency corrective actions are the courses of action that anticipate the occurrence of deficiencies and which are intended to retard the progression of defects. Preventive maintenance is carried out to stave off the inevitable consequences of age and traffic (that is, to prolong the life of the pavement).

This paper describes the steps in the Guidelines, including the distress classification methodology and the procedures leading to selection of the most cost-effective treatment. Illustrative examples of both corrective and preventive measures are used.

Pavement Maintenance Guidelines

The chief aim of the guide is to help our district maintenance field staff identify the pavement distress problem and to choose, from among the available alternatives, the most cost-effective treatment method for that particular distress problem.

Figure 1 shows the various steps in the procedure for maintenance field staff to use the guidelines to make maintenance decisions. The guide includes four essential elements:

(1) a simple way to describe or classify the distress problem,

(2) a list of treatments suitable to correct a particular distress problem, and whether or not patrol forces are adequate for applying the treatment,

(3) a performance standard that tells the men, equipment, and materials needed to do the job—given different traffic conditions and the time required to do it.

(4) a simple method of calculating the comparative unit cost (equivalent annual cost) for each type of treatment to determine which of the suitable alternative treatments is the most cost-effective.

Using the Guidelines for Corrective Maintenance

The Condition Survey

The purpose of the condition survey is to pinpoint the locations on the highway of distresses that require corrective action. At the same time, highway locations where preventive maintenance is needed will be identified. To do this, we have to recognize the distresses present, and we have to classify each problem in simple words. We also want to quantify the problem, that is, we want to be able to say (1) How bad is it? and (2) How big is it? Take this example, alligator cracking...

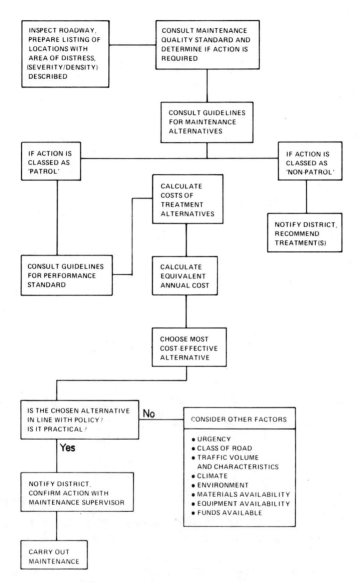

FIG. 1—*Procedures for maintenance staff.*

The guide gives this distress a "name," a description of what it looks like, and a brief summary of reasons why it happens (Fig. 2).

For the first question, "How Bad Is It?" there are three possible answers. They are slight, moderate, and severe. The correct answer may be chosen by simply comparing the observed distress on the road with the descriptions and photos in the guidelines (see Fig. 2).

6 PAVEMENT MAINTENANCE AND REHABILITATION

Description:	Cracks form a network of multi-sided (polygon) blocks resembling the skin of an alligator. The block size can range from 5 to 10 cm to about 50 cm. The alligatored area may or may not be accompanied by distortion in the form of depression, and may occur anywhere on the pavement surface.		
Possible causes:	1. Insufficient pavement strength. 2. Poor base drainage and stiff or brittle asphalt mix at cold temperature.		
Severity:	Class.	**Guidelines** (Base on appearance and surface distortion)	
	Slight	Alligator pattern established with corners of polygon blocks fracturing	
	Moderate	Alligator pattern established with spalling of polygon blocks	
	Severe	Polygon blocks begin to lift; may or may not involve potholes.	
Density:	Local:	Less than 30% of pavement surface affected; distress spotted over localized areas only.	
	General:	More than 30% of pavement surface affected; distress spotted evenly over entire length of pavement section.	

FIG. 2—*Alligator cracking—an extract from the Pavement Maintenance Guide.*

The question, "How Big Is It?" refers to what extent the road surface is covered with alligator cracking. This extent of occurrence is called "density," and it is broken down into two categories, local and general. These two classifications are summarized as follows:

• Class: Guidelines (based on percent of surface area in the pavement section affected by distress).
• Local: Less than 30% of pavement surface area is affected by distress; distress is spotted over localized areas only.
• General: More than 30% of pavement surface area is affected by distress; distress is spotted evenly over the entire length of the pavement section.

Maintenance Performance Standard for Alternative Treatments

Once the alligator cracking has been properly identified and classified according to severity and density, the question arises, "What action?"

The answer is provided in Table 1. Depending on the evaluation of the problem (Col. 1, Table 1), the action may range from "no action" to several possible alternatives. The name of each alternative is given and the reference number of the maintenance performance standard for each is specified (Cols. 2 and 3, Table 1). These performance standards are available in full detail in the guide, and the information on materials, equipment, manpower, methods, and production is used to make a comparative evaluation of the recommended alternatives for cost-effectiveness [1].

We can also decide from the maintenance function classification (Col. 3, Table 1) who is going to carry out the work. Can our patrol do it or is it just too big a job, which must be done either by combined patrol forces, by a special work crew, or by contract? If we decide the job is beyond the resources of our patrol, we submit a "highway maintenance needs report" to our district office (Fig. 3).

Expected Life of Alternative

Obviously, if the safety or minimum comfort level of the travelling public is threatened by a pavement distress, such as a large pothole, something must be done quickly. In short-term emergencies like this, removing or alleviating the problem is all-important, even if it is done expediently and needs to be redone more permanently later on (that is, cost considerations are not the prime factor).

In the medium to long term, however, care must be taken to select the maintenance treatment that is most cost-effective. To evaluate this, we have to know "How long will it last?" Again, the answer is provided by Table 1 (see Col. 4), which lists the expected life of each treatment (based on a survey of maintenance staff with more than 20 years experience).

TABLE 1—*Example of maintenance treatment alternatives—alligator cracking distress.*

Evaluation		Recommended Maintenance Treatment Alternatives	Maintenance Function Classification		Expected Effective Life, Years
Severity	Density		Routine Patrol	Nonpatrol	
Slight	local	no action	...	...	...
	general	no action but monitor closely for future development	...	...	...
Moderate	local	spray patch	1004	...	1
		cold mix patch	1001	...	1
		hot mix patch	1001 1002	1002	4
		hot mix patch for multilanes	...	1002	4
	general	same as above but notify district office of situation and maintain close monitoring	...	...	...
Severe	local	cold mix patch	1001	...	0.5
			1001	1002	...
		hot mix patch	1002	...	3
		excavate, granular, and hot mix patch	1002	1002	7
		improve drainage (additional)	*	*	...
	general	hot mix patch for highways with AADT < 2000 and notify district office	1002	...	2
		mulching for AADT < 2000	...	1014	3
		granular lift and surface treatment for highways with AADT < 2000	...	1017	4
		hot mix patch over selected areas and notify district office for further action for highways with AADT > 2000	1002	1002	3

*Contract only.

Evaluating the Cost-Effectiveness of Alternatives

Our goal in pavement maintenance is to select the best treatment for each situation. By "best," we usually mean the most cost-effective.

Cost comparisons for the alternatives recommended for the particular distress condition are easily done using the maintenance performance standard corresponding to the code numbers of the alternatives given in Table 1 (see Col. 3). The appropriate performance standard gives complete details on the manpower, equipment, and materials needed and the methods to be used for each job, and how much can be done in one day's work. From this we can calculate and compare the cost of each recommended treatment alternative.

The calculation is simple. First, the unit cost of doing the job is

$$\text{unit cost} = \frac{\text{manpower} + \text{equipment} + \text{materials}}{\text{accomplishment per day}}$$

FIG. 3—*Sample of maintenance needs report.*

But knowing the unit cost is not enough to make a good comparison because one treatment may cost the same as another but last twice as long. To take this difference in service life into account, we have to average the cost of a repair over the years until we have to repair it again. This is called the equivalent annual cost, and it is calculated like this

$$\text{equivalent annual cost} = \frac{\text{unit cost}}{\text{expected life of alternative in year(s)}}$$

The treatment with the lowest equivalent annual cost is the most cost-effective of the alternatives and should be carried out unless other factors prevail [1].

Other Factors

The selected treatment, although it is most cost-effective, may sometimes be changed in view of special circumstances. These may be emergencies, class of road, climate, traffic conditions, availability of manpower, funds, materials, and so forth (Fig. 1). It is up to our field staff to consider these factors and to judge whether these "other factors" override cost-effectiveness.

Action Confirmed

Once the best alternative is chosen, the patrol will confirm the action with the district maintenance supervisor, then carry out the work.

Preventive Maintenance

All pavements will benefit from the application of preventive maintenance treatments. However, the beneficial effects will differ depending on the characteristics of the pavement structure. In general, it can be said that the more structurally sound (stronger) pavements would benefit more than pavements that are weaker. The reasoning behind this statement is illustrated in Fig. 4.

The statement can be translated into terms that can be more fully appreciated by financial analysts, that is, the risk that one takes in investing preventive maintenance funds in a stronger pavement is low, while there is a high risk in investing preventive maintenance funds in pavements that are weaker. Thus, one criteria for recognizing candidates for preventive maintenance may be identified by reference to records of pavement condition ratings (PCR). PCR are obtained in condition surveys by a geotechnical engineering staff rather than by a maintenance staff, utilizing somewhat more detailed scales for density and severity of distress than used in the Pavement Maintenance Guidelines, and combining distress with ride quality [3]. The shaded area in Fig. 4, representing candidates for preventive maintenance, shows medium- to low-risk rating pavements, with timing of nonpatrol type preventive maintenance activity occurring between PCR values of 70 and 55.

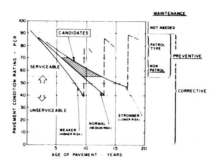

FIG. 4—*Age of pavement versus pavement condition rating showing candidates.*

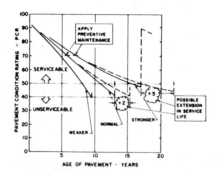

FIG. 5—*Candidates for preventive showing medium- to low-risk rating pavements.*

Flexible Pavement Crack Sealing as Preventive Maintenance

Crack sealing pays off in extending pavement life through waterproofing the pavement surface to prevent softening of the structural layers by water from the surface. Other waterproofing measures, such as sealing granular shoulders and pavement edge drains, are companion preventive maintenance activities. The amount of life extension is uncertain at this time because there are no extensive records, but our belief is based on recently instituted research findings that up to five years extended life is not an unreasonable figure to use [3] (Fig. 5).

Rationale for Crack Sealing

Consider, however, the consequences of not sealing cracks.

It is a well-known fact that cracks in flexible pavement will close and open as seasons change from summer to winter. In Ontario, historical data indicate that this crack movement varies from an average of 3 to 4 mm for longitudinal type of cracks to 7 to 10 mm for transverse type of cracks.

But even in summer when cracks are closed, moisture can enter into the cracks and penetrate to the underlying granular base where it is available to carry fines to the surface under the pumping action of traffic (Fig. 6).

When winter comes, the same cracks open up and water is practically drawn into the cracks, that is, water that has high content of deicing salts (Fig. 7). This salt water subsequently freezes and thaws more times in lower temperatures than pure water would, and this can create "cupping" or "lipping" at the crack (Fig. 8). Cupping or lipping or both contributes significantly to roughness of ride. There is residual roughness even after the spring thaw.

Even more drastic is multiple cracking and spalling that develop into potholes if timely preventive maintenance actions are not taken (Fig. 9). Crack sealing plays a prominent role in preventive maintenance; it seeks to prolong

FIG. 6—*Closed cracks in the summer.*

FIG. 7—*Closed cracks in the winter.*

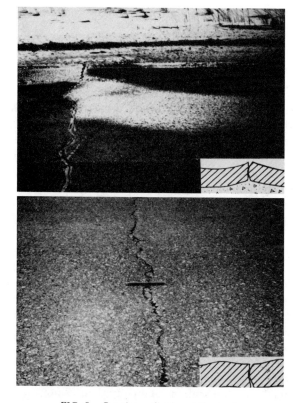

FIG. 8—*Cupping or lipping at the crack.*

FIG. 9—*Multiple cracking and spalling.*

pavement life, mainly to forestall deterioration, by waterproofing of the pavement surface.

Crack Sealing Guidelines

To achieve the extended pavement life we seek, we need appropriate guidelines to ensure the actions we undertake incorporate the most effective materials, methods, and equipment available. Ontario still does not have the definitive answers to sealing cracks in flexible pavements in cold areas, but our experience is sufficient that we were able to formulate trial guidelines for one of our regions (southwestern).

General Policy Statement—Crack rout-and-seal should be considered a preventive maintenance measure with respect to the timing of application. A crack rout-and-seal program should be established from two to five years following construction of every new pavement, hot-mix resurfacing included.

Crack rout-and-seal should not normally be considered a corrective maintenance measure. Do not rout-and-seal cracks if pavement cracking has progressed to the stage where there is more than "one" transverse crack within 10 m (30 ft); rout-and-seal in these conditions would cost as much as a one-lift overlay. Exceptions may be made where crack rout-and-seal is part of a rehabilitation program, or is a measure to delay rehabilitation.

Guidelines for Crack Rout-and-Seal Action are as follows:

1. For flexible pavements, if cracks are (a) Less than 6 mm—do nothing, (b) 6 to 12 mm—rout-and-seal, or (c) greater than 12 mm—clean and seal.

Priority should be given to rout-and-seal of longitudinal single cracks of centerline, midlane, meandering, and wheel track varieties. It is important to catch these at an early stage. Give lower priority to rout-and-seal of transverse cracks unless the cracks are about to develop ravelling, lipping, or cupping. Do not rout-and-seal pavement edge cracks except where pavement shoulders are waterproofed or about to be waterproofed.

2. For composite pavements, if cracks are (a) up to 12 mm—rout-and-seal or (b) greater than 12 mm—clean and seal.

Give priority to rout-and-seal of all cracks where the asphalt surface is less than 125 mm (5 in.) thick over the concrete surface course.

Lower priority may be assigned to rout-and-seal of both longitudinal and transverse cracks where the asphalt surface is greater than 125 mm (5 in.) thick over the concrete surface course. Rout-and-seal only when monitoring shows cracks about to develop ravelling, cupping, or lipping.

Guidelines for Methods and Materials are as follows:

1. Rout size should be 12 mm wide and not less than 10 mm deep, but not to exceed 12 mm.

2. Rout should be thoroughly cleaned and dried (preferably with a hot compressed air lance).

3. A bead of sealant 50 mm wide should be formed over the 12 mm wide rout.

4. Only an approved hot-poured sealant on the designated source list is to be used. All materials must be used within the temperature range specified by the manufacturer.

Need for On-Going Research and Technical Improvement

For the past few years, Ontario has promoted crack sealing as a preventive maintenance measure to prolong pavement life. In the 1982/83 fiscal year, our ministry spent approximately $1.5 million (Canadian) on crack sealing, not including that which was spent by regional and municipal agencies. The question that is again being raised is whether we are getting benefits from this program. When, for whatever reason, the sealant becomes ineffective after only a year or two, this becomes a question worth addressing. It is certainly time to look much more closely at ways to improve the effectiveness of sealing cracks, from both a technical and an operational perspective.

Ontario has, in place, a number of test sections to address these questions [3]. We have also encouraged industry to develop the kind of equipment and materials that will increase quality workmanship and productivity. For example,

1. *Routing*—We are in the process of experimentally altering the configuration of the routing head to obtain sloping rout faces. We are also using a shallow depth of cut that we feel will be more advantageous for successful adhesion. The shallow depth considerably reduces the force applied during extension of the sealant in cold temperatures, and thus reduces the adhesion stresses as well. Adhesion failure has been responsible for the majority of past failures in crack sealing.

Previous observations have shown that overfilled routed cracks (one to one shape) were more successful than underfilled cracks, and that overfilled cracks with a bead shaped with a squeegee were likely to be even more successful.

The shallow-cut end-sloping rout faces enable the sealant to be neatly inset into the pavement, and it does not need to be overfilled with a shaped bead. This lessens labor costs by deleting the extra man required on the squeegee to form the bead. It also reduces material costs as less wastage of sealant will be the norm (Fig. 10).

2. *Cleaning and Drying*—Propane-fuelled hot compressed air lances have been developed and marketed by a number of companies. The hot, high-speed air blows out the cracks and dries up any moisture that might hinder adhesion. Safety problems associated with operators being exposed to prolonged heat have now been alleviated by the operator wearing a pair of wooden clogs in addition to his regular safety shoes/boots (Fig. 11).

3. *Materials*—Softer materials with high extensibility at low temperatures

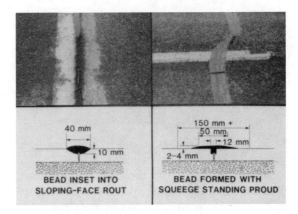

FIG. 10—*Shallow-cut end-sloping rout faces.*

FIG. 11—*Operator wearing a pair of wooden clogs to alleviate problems caused by exposure to prolonged heat.*

are preferable; however, resistance to tire pickup at high temperatures is still needed.

Summary

Our system of determining the what, where, and when of corrective and preventive maintenance, as discussed herein, is really quite straightforward. All the basic facts necessary to make the system work are readily available to our maintenance staff, and they through excellent co-operation with geotechnical staff (who are concerned with pavement design and evaluation) correct pavement deficiencies, the Ontario way.

References

[1] Chong, G. J., Jewer, F., and Macey, K., "Pavement Maintenance Guidelines: Distresses, Maintenance Alternatives and Performance Standards," Report SP-001, revised edition, Ontario Ministry of Transportation and Communications, Downsville, Ontario, 1982.
[2] Phang, W. A. and Blum, W., "Preventive Pavement Maintenance Concepts," Ontario Ministry of Transportation and Communications, Downsville, Ontario, 1980.
[3] Chong, G. J., Phang, W. A., and Wrong, G. A., "Manual for Condition Rating of Flexible Pavements—Distress Manifestations," Report SP-004, Ontario Ministry of Transportation and Communications, Downsville, Ontario, April 1982.
[4] Chong, G. J. and Phang, W. A., "Sealing Cracks in Flexible Pavements in Cold Areas—An Audio Visual Script," Report PAV-83-01, Ontario Ministry of Transportation and Communications, Downsville, Ontario, 1983.

Robert C. Deen,[1] *Herbert F. Southgate,*[1] *and Gary W. Sharpe*[1]

Evaluation of Asphaltic Pavements for Overlay Design

REFERENCE: Deen, R. C., Southgate, H. F., and Sharpe, G. W., "**Evaluation of Asphaltic Pavements for Overlay Design,**" *Pavement Maintenance and Rehabilitation, ASTM STP 881*, B. F. Kallas, Ed., American Society for Testing and Materials, Philadelphia, 1985, pp. 18-32.

ABSTRACT: To evaluate projects involving approximately 200-route-miles of interstate and primary pavements in Kentucky and Tennessee in relatively short time frames, it was decided to test, analyze, and design overlays using test equipment (Road Rater®) and procedures developed by the University of Kentucky Transportation Research Program. This paper presents the analysis methodology and the evaluation and overlay designs for selected projects, including the before and after analysis of milling on one project.

The Road Rater applies a dynamic sinusoidal loading of known force and frequency. The velocity of the vibration waves are measured by sensors and integrated electronically to obtain surface deflections. An analysis of the shape and magnitude of the deflection bowl permits an assessment of whether the structure is performing as anticipated or whether some component is significantly weaker than designed. Analyses permit the determination of the "behavioral" or effective thicknesses of the asphaltic concrete layers and the in-place subgrade moduli. Strip charts of effective thickness and subgrade strength (or alternatively, overlay thicknesses) along the length of a project permit delineation of the project into relatively uniform segments. The arithmetic mean and standard error are determined for each segment to estimate design parameters. The required overlay thickness is the difference between the total thickness required for new construction to carry the anticipated traffic and the behavioral thickness of the existing pavement.

KEY WORDS: pavements, bituminous cements, highways, asphaltic pavements, overlay designs, deflections, deflection bowl, structural condition, nondestructive testing, pavement evaluation, subgrade modulus, asphaltic concrete modulus, equivalent axleloads

The transportation infrastructure, in part consisting of the highway and street networks of this country, is a primary element in economic development and growth. There is now a tremendous burden on highway agencies to

[1]Director, chief research engineer, and chief research engineer, respectively, Kentucky Transportation Research Program, University of Kentucky, 533 South Limestone, Lexington, KY 40506-0043.

protect the financial investment represented by the highway and street systems. To make the most efficient use of available and typically insufficient funds, it is necessary to select and schedule rehabilitation activities on a timely basis.

A number of tools and methodologies are available to the highway engineer and administrator to assist in making decisions as to appropriate rehabilitation strategies. Visual surveys and observations of pavement conditions, measurements of rutting and road roughness (ride quality), measurements of the extent of cracking and patching, and measurements of skid resistance provide input upon which to base decisions for rehabilitation activities. Unfortunately, these approaches of observing or measuring surface manifestations of pavement performance are not always adequate. These traditional procedures may not show imminent, but hidden, structural deterioration.

To assess the structural capacity of a pavement system, pavement deflections (more specifically, dynamic deflections) have been used as an additional input variable to the decision-making process of selecting and scheduling rehabilitation strategies. Dynamic deflections may be induced and measured by such apparatus as the Road Rater®, Dynaflect®, and falling-weight deflectometer. This information, along with the other more conventional input factors of pavement condition and performance, permits a more complete analysis of the sufficiency of pavement systems on a project by project basis; resulting decisions relating to rehabilitation strategies are much more efficient.

The objective of this paper is to summarize and document a methodology that has been developed to evaluate the structural adequacy of asphaltic concrete pavements before preparing overlay designs and recommending other rehabilitation strategies. The methodology is based on elastic theory and a rational pavement thickness design schema. The procedure makes use of dynamic pavement deflections measured by the Road Rater. It also has been demonstrated that pavement deflections obtained with the Dynaflect and adjusted appropriately are compatible with the procedure.

The methodology described herein is founded on a means for calculating theoretically expected deflections, stresses, and strains of a pavement system in such a way as to simulate the sinusoidal loadings to which the pavement is subjected (both by traffic and by testing equipment). This has been done with the use of elastic layer theory as expressed in the Chevron N-layer computer program. Since the behavior of the constituent materials of the pavement system are temperature dependent, a scheme to adjust field measurements of surface deflections to account for temperature distributions within the pavement and seasonal effects is described. Field measurements are matched with theoretical deflections of an appropriately assumed pavement structure. Then a procedure for estimating the structural behavior of existing pavements (in terms of effective subgrade modulus and effective thickness of asphaltic concrete) is presented. Finally, the resulting design parameters and other design inputs are used to estimate thicknesses of asphaltic concrete overlays.

Basis for the Evaluation Methodology

Simulation of Dynamic Deflections by Elastic Theory

Loading—The testing head of the Kentucky Road Rater consists of a vibrating mass of 72.6 kg (160 lb) that impulses the pavement through two feet symmetrically located on either side of a beam that extends ahead and supports the velocity sensors. The forced motion of the pavement is measured by velocity sensors located at 0, 305, 610, and 914 mm (0, 1, 2, and 3 ft) from the center of the test head. When the vibrating mass is lowered to the pavement under a hydraulic pressure of 4.82 MPa (700 lbf/in.^{2}), the static load is 7.43 kN (1670 lbf). At a frequency of 25 Hz and a double amplitude of vibration of 1.52 mm (0.06 in.), the Road Rater has a double-amplitude dynamic force oscillation (about the static load) of 2.67 kN (600 lbf).

The dynamic loading (sine wave) of the Road Rater may be approximated by a square wave such that the maximum value of the square wave is equal to $1/\sqrt{2}$ times the peak value of the sine wave. The maximum and minimum square-wave loadings for the Kentucky Road Rater are 8.37 and 6.49 kN (1882 and 1458 lbf). From symmetry, the maximum and minimum loads on each foot of the test head are equal to 4.19 and 3.24 kN (941 and 729 lbf), respectively.

Input Parameters—Inputs required by the Chevron N-layered program [1] (used to calculate stresses, strains, and deflections) include a contact pressure corresponding to the applied load; the number of layers; and the thickness, Young's modulus, and Poisson's ratio of each layer. The contact pressures of the maximum and minimum loads were selected to maintain the correct area for each loading foot. A range of typical values were used in simulating the Road Rater loadings and deflections.

The modulus of a granular base E_2 is a function of the moduli of the confining layers, that is, the modulus of the asphaltic concrete E_1 and the modulus of the subgrade E_3. Estimation of the modulus of the crushed-stone layer may be determined from the relationship $E_2 = F \times E_3$, where there is an inverse linear relationship between log F and log E_3. The ratio of E_2 to E_3 is equal to 2.8 at a California bearing ratio (CBR) of 7 and to 1 when E_1 equals E_3; that is, $E_1 = E_2 = E_3$ [2,3], the case of a Boussinesq semi-infinite half space. Laboratory triaxial testing also has indicated variations in modulus as a function of confining pressure. A modulus ratio of 2.8 (crushed-stone base to subgrade) at a CBR of 7 represents experience in Kentucky. The modulus of the subgrade (in lbf/in.^{2}) can be approximated by the product of CBR and 1500, a method of estimating moduli adequate for normal design considerations up to a CBR of about 17 to 20 [2-6].

Reference Conditions—The modulus of elasticity of asphaltic concrete varies as a function of frequency of loading and of temperature. Conditions for the historical Kentucky thickness design procedures and the method for conducting Benkelman beam (static) deflection tests correspond to a modulus of 3.31 GPa (480 000 lbf/in.^{2}) at 0.5 Hz and a pavement temperature of 21°C (70°F). At a reference frequency of 25 Hz for the Road Rater, the corresponding modulus of asphaltic concrete at 21°C is 8.27 GPa (1 200 000 lbf/in.^{2}), obtained using Fig. 1, a close approximation of results of laboratory testing by Shook and Kallas [7].

The Deflection Bowl

Analyses of pavement deflections involve examinations of the shapes of deflection bowls [5,6,8-16]. An empirical evaluation of the shape of a deflection

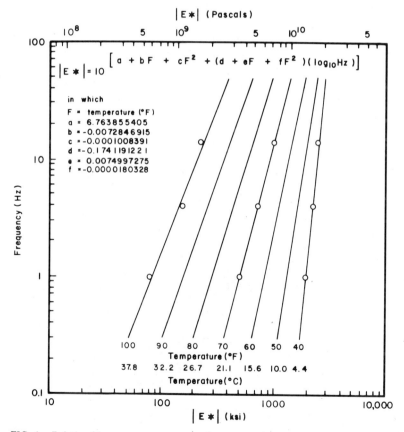

FIG. 1—*Relationships among temperature of pavement, frequency of loading, and modulus of elasticity of the asphaltic concrete.*

bowl involves extrapolating a straight line through the Road Rater deflections of the No. 2 and No. 3 sensors when log deflection is plotted as a function of the arithmetic distance from the load head. The deflection at the position corresponding to the No. 1 sensor is the No. 1 projected deflection (1P in Fig. 2, for example). The slope of the semilog secant line, the difference between the No. 1 projected (1P) and the No. 1 sensor (1M) deflections, and the magnitude of all deflections are all indicative of the shape of the deflection bowl.

Typically, there is a difference between the No. 1 projected and the No. 1 sensor deflections, both for theoretical deflections (calculated using the Chevron N-layered program and design or as-constructed input parameters) and for field-measured deflections. Normally, differences between the No. 1 projected deflection and the No. 1 sensor deflection for both theory and field measurements are the same. However, when these differences are not the same, unanticipated behavior of the pavement system is indicated.

A log-log plot of No. 1 projected deflections versus No. 1 sensor deflections may be used to identify variations in behavior of the pavement structure. The solid line in Fig. 3 shows the theoretical relationship for a given structure and asphaltic concrete modulus. Subgrade modulus increases logarithmically (approximately) along the line as deflections decrease. That approximate logarithmic scale is a function of pavement structure. The zone inside the dashed lines represents an expected variation caused by normal operator "errors" in reading the meters of the Road Rater.

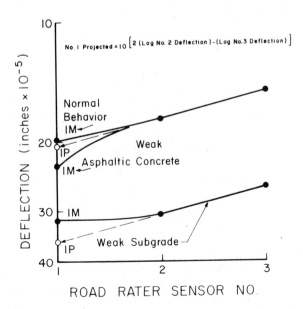

FIG. 2—*Surface deflection as a function of distance from the Road Rater load head, illustrating the determination of the No. 1 projected deflection (1P).*

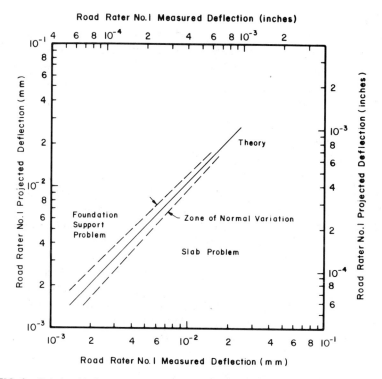

FIG. 3—*Relationship between Road Rater No. 1 projected deflection, and Road Rater No. 1 sensor deflection (measured).*

Effect of Errors and Missing Data

Current procedures utilize deflections of the three sensors nearest the point of load application to evaluate the shape of the deflection bowl. Comparisons between the shape of the measured deflection bowl and the theoretical bowl provide estimates of effective thicknesses and subgrade moduli.

The errors in reading Road Rater meters affect predictions of the behavior of existing pavements. The extent of the effect is a function of the thickness and strength of each layer and the strength of the subgrade. Normal operator error affects more significantly the results of analyses when the magnitudes of deflections are small, because a unit change produces a greater change in predicted subgrade strength or effective pavement thickness or both (see dashed lines in Fig. 3).

Occasionally, situations arise such that data for one of the sensors may be missing or obviously erroneous. In that event, data may be analyzed using procedures published previously [6]. The shortcut procedure reported in this paper is not applicable since that approach is predicated upon an analysis of both the shape and magnitude of the deflection bowl.

When the shortcut analysis is not applicable, it is necessary to use deflections at the sensors for which deflections are not erroneous to estimate the subgrade modulus. A mean subgrade modulus may then be calculated. The deflection for the sensor nearest the point of load application is then used in combination with the estimated subgrade modulus to estimate the effective thickness of asphaltic concrete.

Evaluation Procedures

As-Constructed Thicknesses

To properly evaluate the behavior of asphaltic concrete pavements, thicknesses of the component layers must be determined from the most reliable construction or maintenance records or by coring the pavement, if adequate records are not available. Analysis procedures are predicated on matching measured deflections with some theoretical deflection bowl. There are many combinations of layer thicknesses, layer moduli, and subgrade moduli that may result in a deflection bowl that matches field measurements. Only a few of those combinations represent realistic configurations. Existing layer thicknesses are necessary as a starting point for the analysis and also to assess whether results are realistic.

Adjustments for Nonreference Conditions

Moduli of Asphaltic Concrete—The pavement surface temperature, time of day, and mean air-temperature history for the previous five days are necessary to determine the temperature distributions within the pavement [17,18]. The five-day mean air-temperature history can be obtained from weather records at local offices of the National Oceanic and Atmospheric Administration or local radio and TV stations.

The mean modulus of elasticity of asphaltic concrete is a function of frequency of loading and mean pavement temperature [7,19]. A relationship between modulus and temperature may be developed for the reference frequency of 25 Hz, or any other frequency that may be representative of other dynamic loads (Fig. 1). Thus, a distribution of the modulus through the asphaltic concrete layer for the reference frequency of 25 Hz may be determined for any temperature distribution.

Adjustment Factors for Deflections—Because of the significant effect of temperature on the modulus of elasticity of asphaltic concrete, it is necessary to adjust deflection measurements to a reference temperature and modulus. Relationships shown in Fig. 1 can be used to develop such adjustments. For a given thickness of asphaltic concrete, adjustment factors vary according to changes in the thicknesses of granular base and the values of E_3, but these variations are minimal when compared with variations in adjustment factors for differences in thicknesses of asphaltic concrete layers.

The system to be used to adjust deflections to specific conditions, 25 Hz, a mean pavement temperature of 21°C, and E of 8.27 GPa, was developed using theoretical deflection data corresponding to the No. 1 sensor of the Road Rater. Similar systems also were developed for deflection data for both the No. 2 and No. 3 sensors. Experience has shown that use of a single adjustment factor for all sensors may lead to a skewed deflection bowl that may result in erroneous evaluations. Thus, separate adjustment factors are used for each sensor. Equations representing the relationships have been developed to calculate adjustment factors [20].

It has been shown that there is a variation in predicted subgrade moduli from April to Sept., based on data obtained in Kentucky over a one-year period. Such analyses permit the adjustment of deflection data obtained at any time to equivalent springtime deflections, when the subgrade is typically in the weakest condition. Analyses of Kentucky data indicated that fall tests provide the most consistent long-term indicator of pavement behavior. However, overlay designs are based on the subgrade in its weakest condition. Thus, approximate adjustments of test data to springtime conditions should be made. Tests performed on interstate pavements in Tennessee from Aug. through March confirmed this same pattern. The minimum spring value for Tennessee was approximately 0.55 of the fall value compared to 0.60 for Kentucky.

Evaluation of the Pavement Structure

Estimating Subgrade Strength—For given layer thicknesses, relationships were developed (from elastic theory) between theoretical deflections and subgrade moduli for a constant (reference) asphaltic concrete modulus of elasticity (Fig. 4) [20]. The methodology for using these relationships to estimate subgrade strength has evolved through several stages. The No. 2 and No. 3 deflections are used to compute a No. 1 projected deflection. The measured No. 1 sensor deflection and the No. 1 projected deflection are then plotted and compared to values predicted by elastic theory.

Subgrade moduli may be estimated using deflections, measured by any of the sensors singly or in combination. Moduli may vary slightly, but those variations usually are not significant.

Quantifying Effective Behavior—Measured Road Rater deflection bowls can be evaluated by comparing to theoretically expected relationships. Pavement behavior (or condition) can be given in terms of a predicted subgrade modulus, effective layer thicknesses, and effective moduli of the layers. The effective behavior may be expressed by any combination of these variables that matches the measured deflection bowl. In methodologies presented in this paper, however, pavement behavior is expressed in terms of a predicted subgrade modulus and an effective thickness of "reference" high-quality asphaltic concrete. The effective thickness of the granular base is assumed to be equal to the as-constructed thickness.

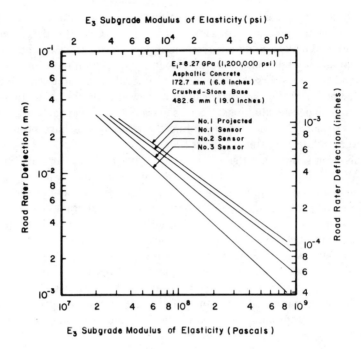

FIG. 4—*Theoretical relationships between Road Rater deflection and subgrade modulus of elasticity for a given pavement structure and asphaltic concrete modulus.*

Determining a "true" and reasonable effective structure of an existing pavement is an iterative process. The reasonableness of the combinations of subgrade strengths and effective thicknesses of the asphaltic concrete is dependent upon the physical constraints (measured deflections and as-constructed thicknesses) of a given pavement structure. The iterative process involves selecting a subgrade modulus and effective thickness and comparing the resulting theoretical deflection bowl to the measured bowl. If the deflection bowls do not match, the subgrade modulus and effective thicknesses are adjusted and the process repeated until a satisfactory match is obtained.

Figure 5 (a combination of Figs. 3 and 4) illustrates a "shortcut" procedure that usually eliminates the need for iterations. The methodology uses the theoretical relationship between No. 1 projected deflections and No. 1 sensor deflections and the theoretical relationship between subgrade moduli of elasticity and No. 1 sensor deflections. A family of lines were constructed to relate theoretical deflection of the No. 1 sensor, subgrade modulus, and thickness of asphaltic concrete [20] as shown on the right side of Fig. 5.

For the Point x in Fig. 5, deflections for the second and third sensors produced a calculated projected deflection too small when compared to the companion measured deflection for the No. 1 sensor. If the measured deflection for the No. 1 sensor is used to estimate the subgrade modulus, the effective

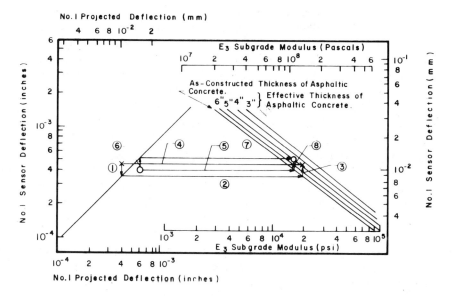

FIG. 5—*Illustration of a method for estimating in-place subgrade modulus and effective thickness of asphaltic concrete.*

behavior is as if the full thickness of the asphaltic concrete were on a weaker subgrade. In that case, the measured deflections for the second and third sensors would not match the theoretical deflections. Therefore, it is necessary to "correct" the "measured" projected deflection to be compatible with deflections at the second and third sensors on the basis of the theoretical relationship between projected deflections and No. 1 sensor deflections (left side of Fig. 5). However, the deflection of the first sensor indicates the thickness of the asphaltic concrete is thinner than the actual thickness. To duplicate the measured deflection bowl, the equivalent structure that matches the condition is one of a thinner asphaltic concrete layer on a stronger subgrade. To obtain that structure having an equivalent behavior, the theoretical deflection is determined by moving vertically from the calculated projected deflection to the solid line on the left side of Fig. 5 (Step 1). Using that point as a turn, move horizontally (Step 2) to obtain the estimated in-place subgrade modulus from the theoretical relationship between deflection and subgrade modulus for the constructed thickness of asphaltic concrete (heavy line on the right side of Fig. 5). Using that estimated subgrade modulus as a turning point, move vertically (Step 3) to the measured deflection (Step 4) for the No. 1 sensor to obtain the estimated thickness of the asphaltic concrete (from the lighter solid lines on the right side of Fig. 5).

The other most commonly measured deflection bowl is illustrated by Point o in Fig. 5. There the deflection bowl is very flat, normally indicating a weak subgrade condition. When deflection bowls of this sort are encountered, the magnitudes of the deflections at the second and third sensors are much larger than theoretically expected and are not compatible with the measured No. 1 deflection. Therefore, it is again necessary to "correct" the measured deflection at the first sensor to be compatible with companion measurements for Sensors 2 and 3. Since the measured deflections at the second and third sensors are indicating a weakened subgrade, the unadjusted measurements are first used to estimate an in-place subgrade strength by moving horizontally along the value of the measured deflection of the No. 1 sensor to the heavy solid line on the right side of Fig. 5 (Step 5). The adjusted deflection for the first sensor (found by moving vertically from the measured deflection to the solid line on the left side of Fig. 5 (Step 6) to locate another turning point) is used to estimate an effective thickness of asphaltic concrete (Steps 7 and 8).

Analyses of field deflections indicated this procedure will produce results that can be used as input into an overlay design process without iteration. Road Rater testing of pavements before and after overlaying shows the ultimate behavior of the overlaid pavement is equal to that of a pavement having a total thickness of reference-quality asphaltic concrete equal to the sum of the effective thickness before overlaying and the overlay thickness [3,21].

Evaluation of Project Data—One method for evaluating the performance of existing pavement structures has been to create a strip chart of overlay thicknesses (determined as indicated in the section entitled Overlay Designs) versus distance along the proposed project. "Uniform" sections of highway may be delineated for possible consideration for various rehabilitation strategies (for example, different overlay designs).

Sampling and Statistics—The sampling interval for deflection testing varies according to specific analysis requirements. The current density of testing for analysis of asphaltic concrete pavements for overlay design purposes is at 0.16-km (0.1-mile) intervals for each direction or lane tested. Generally, overlay thicknesses are not varied in short lengths, and therefore, low-density testing is acceptable. If a specific problem area is to be evaluated, higher densities of testing may be required to delineate the limits of the problem area. In such cases, testing has been done on 30-m (100-ft) intervals or less.

Statistical analyses of the results of the evaluation of deflection data (that is, the required overlay thicknesses) are normally oriented toward the selection of design values. It is desirable to select some level of pavement performance (required overlay thickness) that represents a tolerable balance between some overdesign and some acceptable risk of premature failure. For example, use of mean values for design purposes recognizes a 50% probability of premature failure. Design curves used in Kentucky are based on the 90-percentile level (that is, there is assumed to be only a 10% probability of premature failure). The statistical levels assigned to other aspects of the evaluation of the structural adequacy of pavements (overlay thicknesses) can be varied, depending upon the type of facility under consideration, the funds available for rehabilitation, and the degree of risk that is acceptable.

The larger the sample size, the greater the reliability that may be attributed to the data analysis. A sample size of 30 or more measurements (of deflection bowls) is generally required for most statistics to be considered acceptable, although there are no firm rules regarding sample size. However, the assumption of a normal distribution is more valid with larger sample sizes.

Design Equivalent Axleloads

To prepare an overlay design, it is necessary to estimate or predict the characteristics of the anticipated traffic stream that is to be served by the section of highway under consideration. To use Kentucky's current thickness design procedures, the characteristics of the traffic stream must be expressed in terms of equivalent 80-kN (18-kip) axle loads (EALs) anticipated during the design period. Several procedures are available to obtain such estimates [2,3,22].

Overlay Designs

Once the input parameters (in-place subgrade moduli, effective thicknesses of asphaltic concrete, and design EALs) at each test point have been determined by analyses of deflection and traffic data, overlay designs can be prepared. First, from the design curves, determine the total structural thicknesses for at least three designs using the existing thickness of the crushed-stone base. Plot and connect those points to obtain Curve A in Fig. 6.

For the design EAL, in-place subgrade modulus, and existing thickness of crushed-stone base, determine three total design thicknesses. Plot those designs on Fig. 6 and connect to obtain Curve B. The intersection of Curves A and B is the required total thickness for the design conditions. Overlay requirements of asphaltic concrete can be determined as the difference between the total thickness required for "new" construction (the intersection of Curves A and B) and the effective thickness of the existing pavement at the test point.

Other Applications

The concepts and procedures described in this paper have been applied to dynamic data obtained directly on subgrades, on dense-graded aggregate, on pozzolanic bases, on full-depth asphaltic concrete pavements, and on portland cement concrete pavements. The Chevron N-layered program was used to develop for each case theoretical relationships between deflections and various combinations of layer thicknesses, Poisson's ratios, and moduli. The agreement between the theoretical relationships, Road Rater data, and laboratory data has been amazingly good [23].

The Chevron program also has been used to simulate the Road Rater for

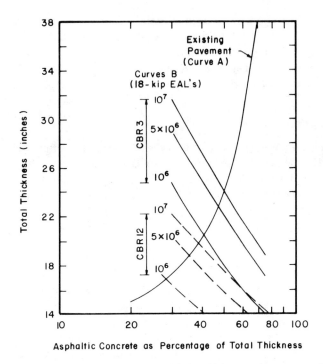

FIG. 6—*Example of relationship between total design thickness and percentage of total thickness caused by asphaltic concrete.*

the analysis of broken and seated portland cement concrete pavements before and after overlaying. In-place subgrade moduli may be estimated from test data obtained on the intact pavement before breaking. Testing after breaking and seating provides estimates of effective moduli of the broken concrete, using the subgrade modulus obtained before breaking. The concepts discussed in this paper are being applied to these situations on an experimental basis. Many questions and relationships still need to be investigated.

Comparisons of sensor deflections from either side of a joint or crack in a portland cement concrete pavement may reveal the effectiveness of load transfer from slab to slab. Procedures utilizing Road Rater measurements for these evaluations are still being studied.

It also has been demonstrated that pavement deflections obtained with the Dynaflect can be analyzed utilizing the concepts presented in this paper. It is necessary, however, to develop relationships among deflections (accounting for sensor locations), moduli (adjusted for frequency of loading), and layer thicknesses that match the dynamic input of the Dynaflect.

The approach presented in this paper greatly simplifies the task for the designer and provides input directly into thickness design nomographs. The methodology offered can be processed by (and was, in fact, developed for) a programmable hand-held calculator. A program has been written for a mainframe computer to process deflection data using the methodology described in this paper. The only advantage of the use of mainframe computers over processing by hand-held calculators is the significant savings in time to reduce the data.

Acknowledgments

The concepts and procedures reported in this paper have been the subject of extensive research in Kentucky over many years. Much of the effort was supported by the Kentucky Transportation Cabinet (and its predecessors); funding has also been provided by the Federal Highway Administration through Highway Planning and Research (HPR) programs. The assistance of the Tennessee Department of Transportation also is acknowledged.

The contents of this report reflect the views of the authors who are responsible for the facts and accuracy of the data presented. The paper does not necessarily reflect the official views or policies of the University of Kentucky or of any of the various supporters of the previous research. This paper does not constitute a standard, specification, or regulation.

References

[1] Michelow, J., "Analyses of Stress and Displacements in N-Layered Elastic System Under a Load Uniformily Distributed on a Circular Area," Chevron Oil Research, Richmond, CA, 24 Sept. 1963.
[2] Havens, J. H., Deen, R. C., and Southgate, H. F., "Design Guide for Bituminous Concrete Pavement Structures," Transportation Research Program, University of Kentucky, Lexington, KY, Aug. 1981.
[3] Southgate, H. F., Deen, R. C., and Havens, J. H., "Development of a Thickness Design System for Bituminous Concrete Pavements," Transportation Research Program, University of Kentucky, Lexington, KY, Nov. 1981.
[4] Witczak, M. W., "Asphalt Pavement Performance at Baltimore-Washington International Airport," The Asphalt Institute, College Park, MD, Research Report 74-2, 1974.
[5] Witczak, M. W., "A Comparison of Layer Theory Design Approaches to Observed Asphalt Airfield Pavement Performance," *Proceedings, The Association of Asphalt Paving Technologists*, Phoenix, AZ, Vol. 44, 1975.
[6] Sharpe, G. W., Southgate, H. F., and Deen, R. C., "Pavement Evaluation Using Dynamic Deflections," Record 700, Transportation Research Board, Washington, DC, 1979.
[7] Shook, J. F. and Kallas, B. F., "Factors Influencing Dynamic Modulus of Asphalt Concrete," *Proceedings, The Association of Asphalt Paving Technologists*, Los Angeles, CA, Vol. 38, 1969.
[8] Kinchen, R. W. and Temple, W. H., "Asphaltic Concrete Overlays of Rigid and Flexible Pavements," Louisiana Department of Transportation and Development, Baton Rouge, LA, Sept. 1977.
[9] Majidzadeh, K., "Dynamic Deflection Study for Pavement Condition Investigation," Ohio Department of Transportation, Columbus, OH, 1974.
[10] Majidzadeh, K., "Pavement Condition Evaluation Utilizing Dynamic Deflection Measurements," Ohio Department of Transportation, Columbus, OH, 1977.
[11] Peterson, G. and Shepherd, L. W., "Deflection Analysis of Flexible Pavements," Utah State Highway Department, Salt Lake City, UT, Jan. 1972.
[12] Peterson, G., "Predicting Performance of Pavements by Deflection," Utah State Highway Department, Salt Lake City, UT, May 1975.

[13] Rufford, P. G., "A Pavement Analysis and Structural Design Procedure Based on Deflection," *Proceedings, Fourth International Conference on the Structural Design of Asphalt Pavements*, University of Michigan, Ann Arbor, MI, 1977.
[14] Southgate, H. F., Sharpe, G. W., and Deen, R. C., "A Rational System for Design of Thickness of Asphalt Concrete Overlays," Record 700, Transportation Research Board, Washington, DC, 1979.
[15] Vaswani, N. K., "Design of Flexible Pavements in Virginia Using AASHITO Road Test Results," Record 291, Highway Research Board, Washington, DC, 1970.
[16] Wiseman, G., Uzan, J., Hoffman, M. S., Ishai, I., and Livneh, M., "Simple Elastic Models for Pavement Evaluation Using Measured Surface Deflections Bowls," *Proceedings, Fourth International Conference on the Structural Design of Asphalt Pavements*, University of Michigan, Ann Arbor, MI, 1977.
[17] Southgate, H. F. and Deen, R. C., "Temperature Distribution Within Asphalt Pavements and Its Relationship to Pavement Deflection," Record 291, Highway Research Board, Washington, DC, 1969.
[18] Southgate, H. F. and Deen, R. C., "Temperature Distribution within Asphalt Pavements," Record 549, Transportation Research Board, Washington, DC, 1975.
[19] Southgate, H. F., Deen, R. C., Havens, J. H., and Drake, W. B., "Kentucky Research: A Flexible Pavement Design and Management System," *Proceedings, Fourth International Conference on the Structural Design of Asphalt Pavements*, University of Michigan, Ann Arbor, MI, 1977.
[20] Southgate, H. F., Sharpe, G. W., Deen, R. C., and Havens, J. H., "Structural Capacity of In-Place Asphaltic Concrete Pavements from Dynamic Deflections," *Proceedings, Fifth International Conference on the Structural Design of Asphalt Pavements*, The Delft University of Technology, Delft, The Netherlands, 1982.
[21] Southgate, H. F., Sharpe, G. W., and Deen, R. C., "Pavement Testing—Before and After Overlay," *Proceedings, International Symposium on Bearing Capacity of Roads and Airfields*, The Norwegian Institute of Technology, Trondheim, Norway, 23-25 June 1982.
[22] Deacon, J. A. and Deen, R. C., "Equivalent Axleloads for Pavement Design," Record 291, Highway Research Board, Washington, DC, 1969.
[23] Sharpe, G. W., Epley, L. E., Allen, D. L., Southgate, H. F., and Deen, R. C., "Low-Strength Pozzolanic Materials for Highway Construction," Transportation Research Program, University of Kentucky, Lexington, KY, Sept. 1983; presented to the Fall Meeting of the American Concrete Institute, Kansas City, MO, Sept. 1983.

Glenn G. Balmer,[1] *Richard A. Zimmer,*[2] *and Richard D. Tonda*[2]

Pavement Friction Measurements and Vehicle Control Reparations for Nontangent Road Sections

REFERENCE: Balmer, G. G., Zimmer, R. A., and Tonda, R. D., **"Pavement Friction Measurements and Vehicle Control Reparations for Nontangent Road Sections,"** *Pavement Maintenance and Rehabilitation, ASTM STP 881*, B. F. Kallas, Ed., American Society for Testing and Materials, Philadelphia, 1985, pp. 33–50.

ABSTRACT: Several measurement systems were evaluated to determine their applicability for measuring wet-pavement friction on transitional roadway sections. It was found that the two-wheel trailer is capable of such measurements if it is instrumented to measure the vertical dynamic test-wheel load as well as the longitudinal drag force during locked-wheel braking. A less desirable approach is to use accelerometers to measure the lateral and vertical accelerations on the trailer and compute the wheel load from these measurements.

There is greater friction deterioration on curves, ramps, intersections, and other transitional roadway areas from traffic wear and the environment than on tangent sections. When these roadway sections become slick or have marginal skid resistance, they can be milled, grooved, resurfaced, or overlaid applying a sprinkle treatment to restore adequate friction characteristics. This paper addresses both the measurement and the restoration aspects of this problem.

KEY WORDS: skid resistance, milling (cold planning), pavement friction, tire-pavement interaction, nontangent road sections, skid number, vehicle control, sprinkle treatment, resurfacing

For many years tire-pavement friction testing on roadways has been conducted in the interest of highway safety. Most of these tests have been on relatively straight level roads, but it is well known there is a greater friction demand on curves or transitions requiring more complex vehicle maneuvers.

---
[1]Highway research engineer, Federal Highway Administration, Turner-Fairbank Highway Research Center, 6300 Georgetown Pike, McLean, VA 22101.
[2]Research instrumentation specialist and assistant research engineer, respectively, Texas Transportation Institute, Texas A&M University, College Station, TX 77843.

The need then exists for measuring surface friction on curves, ramps, intersections, and other transitional roadway areas. Furthermore, remedial procedures should be applied to correct surfaces with low and marginal friction values. This paper addresses both aspects of this problem.

There are a variety of ways to evaluate tire-pavement friction, and a study [1] was conducted to determine which of these methods is possible for measurement of nontangent roadway sections. Two major criteria were used to determine whether an alternate method was worthy of detailed consideration. These were

(1) availability of the device, and its past performance for measurement of friction and

(2) projected performance in a nontangent measurement mode.

Comparisons of Alternative Measurement Methods

To facilitate comparisons of measurement methods, the following principles were selected as equipment guidelines for measuring wet tire-pavement friction for road segments other than tangents:

1. The equipment shall be an efficient measurement system for road surveys at traffic speeds.
2. It shall not interfere with traffic flow or at most cause minor interference.
3. It shall be rugged and require minimum maintenance.
4. The initial and operating costs shall be low for use by state highway agencies.
5. The equipment shall not be complex but be readily operatable by highway technicians.
6. It shall be designed for wet-pavement testing, have its own watering system, and be capable of testing under all weather conditions, except possibly during freezing temperatures.
7. It shall contain an efficient data collection and processing system.
8. Simple, feasible calibration procedures shall be readily available.
9. The equipment shall have an existing capability for measuring nontangent sections, albeit that ability may be severely constrained in some instances.
10. It shall be adaptable to more extreme nontangent sections.
11. There shall be some ability to correlate the equipment output to vehicle behavior.

Table 1 lists evaluations of the equipment considered during this study. These are based on the eleven principles above and scored on a basis of ten. Principles nine and ten are double weighted because of their importance.

TABLE 1—*Comparisons of alternative measurement methods.*

Mu-Meter	DBV[a]	Two-Wheel Trailer	Single-Wheel Trailer	Saab Vehicle	BPT[b]	Evaluation Criteria
10	4	10	10	10	2	efficient for road surveys
10	2	10	10	10	1	noninterference with traffic flow
5	7	8	7	7	9	rugged (minimum maintenance)
4	7	1	2	4	9	low initial and operating costs
6	8	6	6	5	10	simplicity
10	5	10	10	10	1	watering system
8	5	8	8	8	5	efficient data system
8	7	8	7	5	10	readily calibrated
1	3	5	4	6	1	capable for nontangent sections ($w \times 2$)
4	4	8	6	7	1	adaptable to extreme nontangent sections ($w \times 2$)
4	9	8	8	9	1	correlates with vehicle behavior
69	68	95	88	94	52	Totals

[a] DBV is diagonal braked vehicle.
[b] BPT is British pendulum tester.

Equipment Investigation

Testing was conducted with available equipment to further evaluate their measurement capabilities for transient roadway sections. If the tests in a curve on a friction surface, with no appreciable grade or superelevation, were successful, comparison tests were run in a straight course on the same surface. Tests were later conducted also on the roadway with some of the equipment.

Mu-Meter

A Mu-Meter was prepared for testing in accordance with the ASTM Test Method for Side Force Friction on Paved Surfaces Using the Mu-Meter (E 670) and towed with a pickup truck. The Mu-Meter is a small trailer-type unit that measures the side force friction generated between the test surface and the toed-out smooth tires. The friction values are continuously recorded and reported as Mu-numbers, Mu-N.

The results of this testing are shown [1] in Fig. 1. Faulty test limits are indicated by ± 5 Mu-N lines. It is readily apparent that the Mu-Meter data become unreliable at a relatively low lateral acceleration level.

This effect is due to the nature of the Mu-Meter that relies on the vertical loads to be the same on both the left and the right toed-out wheels. As the left

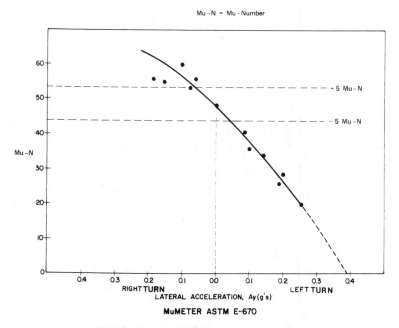

FIG. 1—*Results of Mu-Meter testing program.*

wheel is attached to the pivoting frame member and the right is attached to the remainder of the unit a nonsymmetric system exists. As the unit traverses a left curve, the left wheel is vertically unloaded, which reduces the tensile force on the measuring cell. In a right curve, the left wheel vertical load is increased without proportionally decreasing the restraining force, thus the Mu-numbers increase.

Clearly the Mu-Meter is not satisfactory for measurements on a curve.

Diagonal Braked Vehicle (DBV)

This testing device consists of a passenger car with a diagonal pair of wheels braked for test purposes, in accordance with ASTM Measurement of Skid Resistance on Paved Surfaces Using a Passenger Vehicle Diagonal Braking Technique (E 503). The test pavement is normally wetted by another vehicle just before the test run. A deceleration increment may be used in lieu of measuring the stopping distance.

A limited series of runs were conducted, and the results are shown in Fig. 2. The level of stability was first determined by conducting tests at increasing levels of lateral acceleration by varying the speed through a fixed radius curve.

The same pair of wheels were locked in both the right- and left-turn testing, left, front, and right rear. The stopping distance number (SDN) shows good

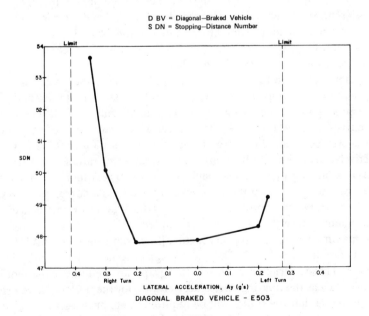

FIG. 2—*Results of DBV testing program.*

consistency up to about 0.2 g, where it starts to diverge. This was apparently because of the increasing vehicle yaw, developing side force on the rolling wheels. This condition was detected by the driver, who had to make steering corrections to stay within the limits of the curve.

Based on these tests and published information, no practical application was found for the conventional DBV in most nontangent measurements. However, the DBV could be used in special cases such as T-type intersections.

Two-Wheel Trailer

In order to quantify the testing limitations of a friction trailer than conforms to American Association of State Highway and Transportation Officials (AASHTO) Standard Method of Test for Frictional Properties of Paved Surfaces Using a Full-Scale Tire (T 242) (equivalent to ASTM Test Method for Skid Resistance of Paved Surfaces Using a Full-Scale Tire [E 274]), physical tests were conducted with it. At least one wheel of the trailer has the capability for remote locked-wheel braking. Simultaneously a controlled amount of water is delivered in front of the wheel for wet-pavement testing. The trailer is instrumented to measure the longitudinal drag force during braking, and the friction number FN or skid number SN is the ratio of the drag force to the test-wheel load multiplied by 100. The wheel load can be determined from the static weight or measured during testing. The latter has been found to be desirable for nontangent testing.

The left wheel, the test wheel of the trailer employed predominantly during this research, contained a biaxial transducer. It was, therefore, not only capable of measuring the longitudinal force, but also the dynamic load on the test wheel when testing, even on a curve or during other vehicle maneuvers. This yielded the instantaneous friction number.

Attention was given to testing on a curve because of the potentially high centrifugal forces acting at the trailer center of mass. Testing was conducted on an asphalt pad, a concrete pad, and a Jennite pad; all were flat. On the test pads, curves were delineated on the surfaces with a radius of 107 m (350 ft). These curves were chosen to produce lateral acceleration of 0.3 and 0.2 g, respectively, at 65 km/h (40 mph). The lateral acceleration could then be varied by adjusting the speed through the curve. It was found that the tow vehicle is more stable than the trailer when testing on a curve, and the safe limit of the trailer is approximately 0.33-g lateral acceleration; hence, few measurements above that limit were attempted.

The maximum safe side friction factors [2] recommended by AASHO for roadways are shown in Fig. 3. These are design criteria for the construction of roadways. The side friction factors of this graph are well under the requirement of 0.33 g as the limiting lateral acceleration for the trailer. Therefore, it is not probable that the trailer will be required to conduct tests on highways that will produce lateral accelerations as great as 0.33 g.

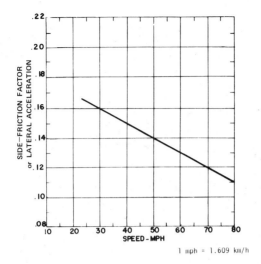

FIG. 3—*AASHO maximum safe-side friction factor.*

Figure 4 shows the test results on the asphalt pad; each data point is the average of six runs. The 2 standard deviation (2 *SD*) limits and the 5 skid number (5 *SN*) limits are shown on the graph. Similar data for concrete and Jennite are given in Figs. 5 and 6. All of the data points except 4 fell within 2 *SD*, and three of these were at 0.3-*g* lateral acceleration.

The stability or instability of a trailer on a curve is directly associated with the lateral acceleration acting on the center of gravity. When the trailer test wheel is locked, that wheel loses virtually all of its side force restraint. Therefore, the unlocked wheel of the trailer must provide the side force resistance to constrain the lateral acceleration of the trailer when testing on a curve, if the trailer is to be stable.

Side Force Measurement Transducer

To further investigate the forces acting on the trailer, a triaxial transducer was installed on the unbraked, right trailer wheel. It was fitted by means of adaptor plates.

The triaxial transducer is a three-force, three-moment transducer. It was used (1) for verification of the computer simulation presented in Ref *1* and (2) to provide better quantifiable measures of trailer limitations. The platform described in ASTM Calibrating a Wheel Force or Torque Transducer Using a Calibration Platform (User Level) (E 556) was used to calibrate the three force axes.

Typical side force data derived from the transducer during a lockup of the left trailer wheel, while negotiating a 20° curve at 48 km/h (30 mph), are

40 PAVEMENT MAINTENANCE AND REHABILITATION

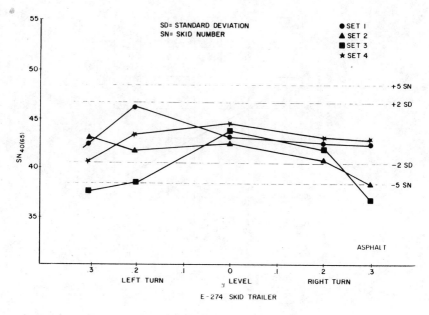

FIG. 4—*Curvature effects on asphalt pavement.*

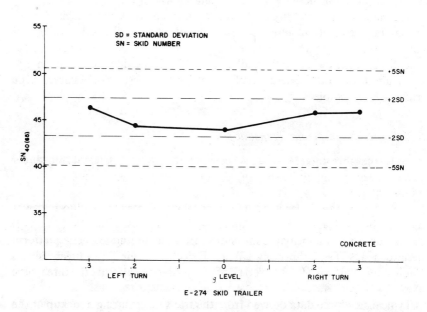

FIG. 5—*Curvature effects on concrete.*

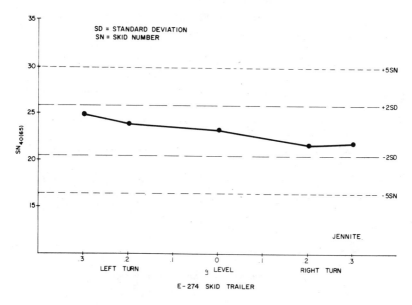

FIG. 6—*Curvature effects on jennite.*

shown in Fig. 7 in the upper curve. This illustrates the magnitudes of the side force on the right wheel before, during, and subsequent to braking the left wheel. The lower trace is an example of the side force developed even in a normal straight-line friction test. This side force is caused by the test wheel tending to align itself directly behind the hitch point during the lockup period. It is resisted by the lateral friction of the free-rolling wheel.

The test results show that the free-rolling wheel of the two-wheel trailer provides the necessary side force resistance during tangent and nontangent testing to maintain trailer stability.

Articulation Angle

To determine the amount of articulation angle change that occurred when the left trailer wheel was braked, a rotational potentiometer was attached between the tow vehicle and the test trailer by means of flexible shaft. This instrumentation allowed measurements in the yaw plane only. The output of the transducer was conditioned and placed on a strip chart recorder for evaluation. This measurement also acted as an indicator of the incipient trailer slip out to the tow vehicle driver.

An example output from these measurements is shown in Table 2 under Columns 3 and 5, the articulation angle. These figures represent the differences between the steady state turn, wheel locked, and unlocked conditions.

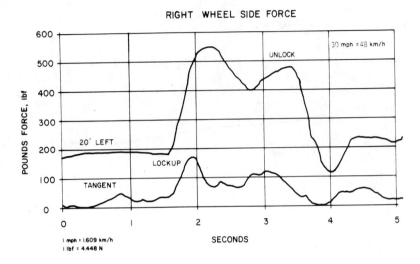

FIG. 7—*Right-wheel side force during lockup.*

TABLE 2—*Curvature testing results.*[a]

Turns	Run #1 TTI FN	Run #1 Articulation Angle, degrees	Run #2 TTI FN	Run #2 Articulation Angle, degrees	ASSMS FN
Left turn area, tangent	42.5 $\sigma = 1.72$	+0.4	42.9 $\sigma = 1.26$	+0.4	42.4 $\sigma = 1.61$
Right turn area, tangent	43.1 $\sigma = 2.10$	+0.5	41.6 $\sigma = 1.90$	+0.3	44.5 $\sigma = 1.01$
Left turn, 65 km/h (40 mph) Radius = 162 m (532 ft), 0.20 g	46.3 $\sigma = 0.91$	+1.2	41.8 $\sigma = 1.85$	+1.4	38.5 $\sigma = 1.08$
Right turn, 65 km/h (40 mph) Radius = 162 m (532 ft), 0.20 g	44.5 $\sigma = 1.03$	0.0	40.9 $\sigma = 1.88$	0.0	42.0 $\sigma = 1.65$
Left turn, 58 km/h (36 mph) Radius = 107 m (351 ft), 0.25 g	44.5 $\sigma = 1.58$	+1.7	41.9 $\sigma = 1.67$	+1.5	41.9 $\sigma = 1.52$
Right turn, 58 km/h (36 mph) Radius = 107 m (351 ft), 0.25 g	38.0 $\sigma = 2.13$	−0.3	39.1 $\sigma = 2.36$	+0.2	39.1 $\sigma = 1.44$
Left turn, 65 km/h (40 mph) Radius = 107 m (351 ft), 0.30 g	46.3 $\sigma = 0.86$	+2.6	40.3 $\sigma = 3.24$	+2.2	37.7 $\sigma = 1.12$
Right turn, 65 km/h (40 mph) Radius = 107 m (351 ft), 0.30 g	42.4 $\sigma = 0.68$	−5.0[b]	38.2 $\sigma = 0.77$	−0.6	36.7 $\sigma = 1.27$

[a] Each condition consists of six runs. The symbol σ is the standard error of the measurements.
[b] No equilibrium was reached; the articulation angle continued to increase during locked-wheel braking.

Notice that an angular shift occurred even in a tangent wheel lock. This is due to the aligning moment produced as the test wheel exerts drag.

Friction tests were conducted both with the Texas Transportation Institute (TTI) trailer and with the Area Reference Skid Measurement System (ARSMS) trailer. Table 2 shows that the friction numbers obtained by the two trailers are comparable, and the articulation angle is small for both tangent measurements and measurements on the curves. This means that the trailers were relatively stable during testing both on the tangent and on the curves.

Testing at Intersections

Friction tests can be conducted with a two-wheel trailer at roadway intersections or approaches to intersections. However, traffic would need to be prohibited or routed through a detour during such testing to avoid conflict.

T-type intersections, abrupt right or left turns, and other special areas can not be tested at the conventional test speed with a two-wheel trailer because there is not sufficient area for maneuvering or stopping. In order to quantify the limitation of the friction trailer for such locations, physical tests were conducted on a large asphalt paved surface. A worst-case, T-type intersection was delineated near the center of the surface. Thus, extreme braking and acceleration maneuvers could be executed with ample recovery area for the safety of the driver and equipment. The test program showed that accurate, repeatable friction values can be obtained at distances exceeding 105 m (344 ft) from the intersection for a speed of 65 km/h (40 mph), exceeding 40 m (131 ft) for 32 km/h (20 mph), and exceeding 22 m (72 ft) for 16 km/h (10 mph). The friction number corresponding to the decreasing test speeds above were 43, 46, and 51, respectively.

Dynamic Vertical Test-Wheel Load

The research presented in Ref *1* and discussed in this paper shows that the two-wheel trailer is capable of measuring the friction characteristics on nontangent roadway sections when the trailer is instrumented to measure both the dynamic vertical test-wheel load and the longitudinal drag force during locked-wheel braking.

A second-best approach was devised for trailers that do not measure the dynamic wheel load. A device was developed [*1*] to provide a continuous readout of test-wheel load by measuring the forces acting on the trailer center of gravity (c.g.). Two accelerometers were used to measure the lateral and vertical accelerations of the trailer and provide the capability of indicating forces acting at the c.g. by applying

$$F = ma$$

where

F = force, N (lbf),
m = mass, kg (lb), and
a = acceleration, g.

By knowing the forces acting at the trailer's c.g. and by knowing the mass and geometrics of the given trailer, it is possible to compute the change in vertical load at the test tire. The following equation applied for this purpose, though not perfect, provides a close approximation to the test-wheel load

$$W = A_Y W_T (h/t) + A_Z W_L$$

where

W = dynamic-vertical load on the test wheel, N (lbf),
A_Y = lateral acceleration, g,
W_T = static-vertical weight of the test trailer, N (lbf),
h = height of c.g., mm (in.),
t = wheel track, mm (in.),
A_Z = vertical acceleration, g, and
W_L = static-vertical load of the test tire, N (lbf).

The c.g. of the trailer was determined by test method Society of Automotive Engineers (SAE) Center of Gravity Test Code (J 874a).

Inertial Wheel-Load Computation Unit

A real-time computation system was developed using an analog computer with common operational amplifiers, as the above equation is linear. The only variables that need to be measured are the lateral and vertical accelerations, as the other quantities are held fixed and are set in the gain elements of the circuit.

The details of this computation system are given in Ref 3, and the results, Fig. 8, from the inertial wheel-load computation unit (dashed line) are compared with the analog output from the wheel-vertical-load transducer (solid line).

In this paper, it is sufficient to state that approximate results can be computed for the dynamic vertical test-wheel load from the measurements of the two accelerometers.

Single-Wheel Tester

The Pennsylvania Transportation Institute (PTI) Mark III Pavement Friction Tester is a single-wheel trailer capable of operating in the locked-wheel,

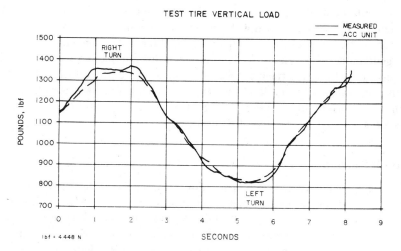

FIG. 8—*Computed versus measured test-wheel load (track).*

transient-slip, and yaw modes. The single-wheel tester uses a hydraulic cylinder to force the wheel into a yaw angle up to 12° from the direction of travel. The measurement system consists of a three-force and three-moment measuring hub. The towing vehicle contains an on-board watering system for wet-pavement testing. The equipment has the associated data collection and processing electronics housed in the tow vehicle to gather and store testing results.

It was not feasible to obtain a unit and test it during this research, but a study of the equipment revealed that it should be quite capable for testing nontangent sections. The acceptable level of lateral force could be somewhat higher than for the two-wheel trailer, because the tow vehicle maintains the stability of the testing trailer rather than depending upon the one rolling wheel of a two-wheel trailer.

There are only a few single-wheel trailers used for friction measurements in the United States; whereas, many of the states own two-wheeled trailers. Testing during this research has shown that the two-wheel trailer is capable of testing nontangent sections. These are the principal reasons for recommending the two-wheel trailer for nontangent friction measurements rather than the single-wheel trailer. Similar statements apply to the Saab friction tester discussed in the next section.

Saab Friction Tester

The Saab friction tester is a specially equipped vehicle with the measuring wheel located between the rear tires, and driven from the differential at a constant 15% slip relative to the drive wheels. It is conceivable that modifications could be made to provide complete wheel lockup during braking. The system contains an on-board watering system operated by a pump.

The system appears to be capable of making measurements in a nontangent mode of operation as it is basically an automobile with a lightly loaded test wheel underneath. Based on the literature, the vehicle should be able to operate at higher g-level conditions than a tow-vehicle test-trailer combination. Correlation between the Saab friction tester and the locked-wheel trailer is not well established at this time.

British Pendulum Tester

The British pendulum tester is a stationary, impact type device for measuring the energy loss when a rubber slider edge is propelled over a test surface. This laboratory device is portable and is capable of making various field surface friction measurements. It is operated in a static condition, in that the instrument does not travel along the roadway as a vehicle does, but is placed in position by an operator. It is necessary to barricade the roadway or route the traffic through a detour during pendulum tests on roadways. It is only operable on a relatively level surface, which eliminates most superelevations, crowns, and grades; although this difficulty might possibly be overcome. The pendulum is most sensitive to pavement microtexture and less sensitive to macrotexture; therefore, the measurements do not correlate well with trailer test results.

The device was considered not as a primary measurement system, because of its obvious deficiencies relative to items 1, 2, and 6 in Table 1 and other shortcomings, but the pendulum can augment the primary system by providing data where it would be difficult or impossible to operate an in-motion vehicle-type measurement system.

Remedial Procedures to Improve Friction Characteristics

Measurement of pavement friction only is not sufficient. If the pavement becomes slick or marginal, it is expedient to apply procedures to improve or restore the friction characteristics. This is especially critical for nontangent sections. Several methods of improving skid resistance and vehicle control will be discussed in the following paragraphs.

Milling or Cold Planing

Cold-planing machines (such as a Roto-Mill) equipped with tungsten-carbide-tipped teeth as cutting edges are capable of removing the surface of pavements. Machines of this nature will leave a skid resistance surface after being milled and will, in addition, decrease the road roughness. The test results in Table 3 from Iowa [4] show test values before and after milling.

TABLE 3—*Roto-milling test results.*

Parameters	Before	After
Skid resistance		
SN_{30}	44	69
	47	69
SN_{40}	38	60
	40	59
	36	58
	31	57
SN_{50}	29	52
	32	54
Road roughness	before	after
Profile indices, in./mile[a]	68.1	12.1
	66.5	10.1
	25.0	7.4
	25.5	4.6

[a] 1 in. = 25.4 mm and 1 mile = 1.609 km.

Cold-planing equipment has been used extensively and very successfully in removing asphalt surfaces from rutted and worn pavements. This procedure is most significant in improving the safety characteristics and the riding quality of pavements. The equipment is less successful for portland cement concrete (PCC), because the machine may cause spalling at cracks and joints; however, it improves the skid resistance.

The milling can be used for leveling and scarifying the pavement before resurfacing. This procedure prepares the pavement for better bonding, and the material removed from the pavement can be recycled in overlays and reconstruction. It can be used in new or reconstructed pavements in the subbase, base, or to build the shoulders.

The pavement surface can similarly be removed and leveled with diamond saw equipment.

Grooving

Hardened pavements can be grooved with diamond saws. Common highway groove width, depth, and pitch patterns are

2.4 by 4.8 by 19 mm (0.095 by 3/16 by 3/4 in.) and
3.2 by 3.2 by 19 mm (1/8 by 1/8 by 3/4 in.).

Before and after accident statistics [5] show that grooving has been very effective in reducing vehicle accidents.

Longitudinal grooves will improve vehicle stability and control on curves and ramps, because the vehicle tends to follow the longitudinal grooves. They will also decrease hydroplaning, as the grooves provide escape channels for the water in wet weather.

Transverse grooves [6] are desirable for approaches to intersection, toll booths, or areas where stopping distance is critical. The transverse grooves allow shorter braking distances, because the grooves produce sharp edges to grip the tires. This is especially important on wet pavements. The transverse grooves accommodate shorter escape channels for the water, thus they are even more effective in reducing hydroplaning than longitudinal grooves.

Grooving is an excellent remedial treatment for nontangent sections of roadways that have become slick, as vehicle control and stopping distance are more critical than on tangent sections. Vehicles are often traveling at high speeds along tangent sections, and the drivers may at times fail to properly reduce their speed before entering a transition; therefore, the friction characteristics of the nontangent sections should be above average.

Grooving is not durable in most new or soft asphalt pavements, but it may be used on oxidized bituminous roadways that have become brittle and on PCC pavements.

Resurfacing

Conventional resurfacing is another excellent method of improving the skid resistance of nontangent sections of roadways; however, it is usually more expensive than milling or grooving.

It is requisite that skid resistant aggregates be used in the overlay to provide adequate friction characteristics. It may be necessary to only resurface egress and ingress ramps, curves, or approaches to intersections if the tangent sections of the roadway have satisfactory skid resistance.

It may be desirable to mill or scarify the pavement before resurfacing to level it and obtain better bond between the pavement and the overlay.

Sprinkle Treatment

The skid resistance of an overlay or of a new pavement can be enhanced by applying a sprinkle treatment [7], especially if the aggregates in the mix are susceptible to polishing readily. This treatment can be applied when critical areas such as egress ramps, curves, or approaches to intersections are resurfaced. In this paper, it is sufficient to state that an adequate amount of high quality aggregate be properly added to the surface during construction if the friction characteristics are to be markedly improved.

Open-Graded Asphalt Friction Course (OGAFC)

An OGAFC [8] is desirable as an overlay to develop better tire-pavement contact during wet weather. The open spaces in the pavement surface allow better expulsion of the water between the tires and the roadway. This decreases the hydroplaning tendencies and improves tire-pavement interactions. Debris from pavement wear and infiltration may lodge in the spaces after a period of time. This will render the OGAFC less effective.

Seal and Chip Coats

A seal coat with chips applied to the pavement surface will also improve the friction characteristics; however, traffic will tend to erode the chips if they are not well bonded. It is essential that the chips be harsh and durable to provide enduring skid resistance.

Conclusions

1. Pavement friction measurements can be competently conducted on nontangent as well as tangent roadway sections.
2. Dependable test results are obtainable with a two-wheel trailer instrumented to measure both the dynamic vertical test-wheel load and the longitudinal drag force.
3. As a second-best alternative, the dynamic wheel load can be computed from accelerometer-measured lateral and vertical accelerations of the two-wheel trailer. This procedure is an approach that can be used for trailers that do not have dynamic-wheel-load measurement capability.
4. Neither the Mu-Meter nor the diagonal braked vehicle is satisfactory for pavement friction testing on most nontangent road sections.
5. Study shows that either a single-wheel friction trailer or the Saab friction tester could be used to conduct nontangent-roadway, pavement-friction measurements.
6. Milling is a practical method of restoring skid resistance to roadways, including nontangent sections, with low or marginal friction characteristics.
7. Pavement grooving is an effective means of significantly reducing vehicle accidents on roadways. Longitudinal grooves will increase vehicle stability and control on curves, and transverse grooves will allow better braking distances in critical stopping areas.
8. Resurfacing is a procedure for a longer lasting improvement in friction characteristics; furthermore it will reduce road roughness and strengthen the pavement. Sprinkle treatments will enhance pavement skid resistance, and an open-graded asphalt friction course will provide an excellent skid-resistance surface if quality aggregates are used in the mix.

References

[1] Zimmer, R. A. and Tonda, R. D., *Pavement Friction Measurements on Nontangent Sections of Roadways*, Vol. 2, Report FHWA/RD-82/150, Federal Highway Administration, McLean, VA, October 1983, pp. 1–94.

[2] *A Policy of Geometric Design of Rural Highways*, American Association of State Highway Officials, Washington, DC, 1965, p. 156.
[3] Zimmer, R. A. and Tonda, R. D., *Pavement Friction Measurements on Nontangent Sections of Roadways*, Vol. 3, Report FHWA/RD-82/151, Federal Highway Administration, McLean, VA, Oct. 1983, pp. 1-90.
[4] Schroeder, C. J. and Bergren, J. V., *An Evaluation of the Roto-Mill Profiler on Concrete Pavements in Iowa*, Iowa Department of Transportation, Ames, IA, Nov. 1976, pp. 1-4.
[5] Beaton, J. L., Zube, E., and Skog, J., *Reduction of Accidents by Pavement Grooving*, Special Report 101, Highway Research Board, Washington, DC, 1969, pp. 110-125.
[6] Balmer, G. G. and Gallaway, B. M., *Frictional Interaction of Tire and Pavement, STP 793*, "Pavement Design and Controls for Minimizing Automotive Hydroplaning and Increasing Traction," American Society for Testing and Materials, Philadelphia, 1983, pp. 167-190.
[7] Weber, H. H., Jr., *Sprinkle Treatment of Asphalt Surfaces State-of-the-Art Review*, Interim Report FHWA-DP-50-13, Federal Highway Administration, Washington, DC, Nov. 1982, pp. 1-96.
[8] Smith, R. W., Rice, J. M., and Spelman, S. R., "Design of Open-Graded Asphalt Friction Courses," Report FHWA-RD-74-2, Federal Highway Administration, Washington, DC, Jan. 1974, pp. 1-38.

Norman E. Knight[1] and Gary L. Hoffman[1]

Heavy Duty Membrane for the Reduction of Reflective Cracking in Bituminous Concrete Overlays

REFERENCE: Knight, N. E. and Hoffman, G. L., "**Heavy Duty Membrane for the Reduction of Reflective Cracking in Bituminous Concrete Overlays,**" *Pavement Maintenance and Rehabilitation, ASTM STP 881*, B. F. Kallas, Ed., American Society for Testing and Materials, Philadelphia, 1985, pp. 51-64.

ABSTRACT: The prevalence of reflective cracking in asphaltic concrete overlays is a major factor contributing to the premature failure in the pavement system. This reflective cracking is caused by cyclic stresses induced in the overlay by movements in the underlying pavement. Recent work done with heavy duty membranes has shown that they may be useful in retarding this reflective crack formation. Seven different types of heavy duty membranes were placed over portland cement concrete pavement joints at a site in Pennsylvania before the roadway was overlayed with asphaltic concrete. Control sections, without any membranes, were also built into the project for comparison purposes. This work will evaluate the ability of these membranes to reduce the occurrence of reflective cracking over transverse and longitudinal joints and to function as a water stop once cracking has occurred.

KEY WORDS: bituminous concretes, pavements, cracks, heavy duty membranes, reflective cracking, asphaltic overlays

The prevalence of cracking in bituminous concrete over overlays is a major factor contributing to the premature failure of the pavement system. This cracking is caused by cyclic stresses induced in the overlay by movement, whether vertical or horizontal, in the underlying pavement. These cracks permit water to enter the pavement structure, promote raveling and secondary shear cracking of the bituminous pavement, and present a continuous maintenance problem. Damage induced by these cracks ultimately requires extensive repairs or a complete overlay even though intermediate areas of the pave-

---

[1]Materials engineer and director, respectively, Bureau of Bridge and Roadway Technology, Pennsylvania Department of Transportation, Room 1009 Transportation and Safety Building, Harrisburg, PA 17120.

ment surface may have many years of remaining serviceable life. Thus, any method of preventing the formation or reducing the severity of cracks becomes highly desirable.

Recent work [1] done with the use of heavy duty membranes to reduce reflective cracking has shown good success. Subsequently, seven different types of heavy duty membranes were placed over reinforced cement concrete pavement (RCCP) joints at one site in Pennsylvania before the roadway was overlayed with bituminous concrete. Control sections, without any membranes, were also built into the project for comparison purposes.

The objective of this study was to evaluate the effectiveness of these heavy duty membrane materials under Pennsylvania's climatological conditions. The ability of these membranes to prevent or reduce the occurrence of reflective cracking in bituminous concrete over transverse and longitudinal joints and cracks in RCCP was determined. These materials were observed over two freezing seasons, and the number and length of reflective cracks in each section were specifically determined.

Project Site Data

The project site is Traffic Route 422 (Legislative Route [L.R.] 149, Section 22M) in Sinking Spring Borough, Berks County, PA (Fig. 1). This highway is designated Class II, has an average daily traffic (ADT) of 17 760, and is located in an urban area that has curbing and storm drains. The existing reinforced concrete pavement has slab lengths of approximately 23.5 m (77 ft); most of the slabs in the state are 19 m (62.5 ft) in length. This concrete pavement has three 3.4-m (11-ft) wide paving lanes and two 2.3-m (7½-ft) wide concrete shoulders, which resulted in 14.6-m (48-ft) curb to curb total width.

The existing longitudinal and transverse joints were filled with fine aggregate debris and embrittled PA "J-1" asphaltic joint sealing material. Some minor spalling, but no faulting, existed along these joints. The transverse joint widths varied from 19.1 to 31.8 mm (¾ to 1¼ in.), and the two longitudinal joints varied in width from 6.4 to 12.7 mm (¼ to ½ in.).

Materials

The membranes used are classed by the manufacturers as heavy duty membranes; a description of these products follows:

Section 1, Polyguard 665®—This membrane is manufactured by Polyguard Products, Inc., of Prior, OK, and consists of a rubberized asphalt waterproofing layer with a polypropylene woven mesh laminated to the top surface. This product is 1.6 mm (65 mil) thick, 305 mm (12 in.) wide, has a roll length of 60.5 m (200 ft), and has a plastic release film on the underside to protect the bottom asphaltic layer until installed. Polyguard requires the application of Polyguard 650 primer at the rate of 5 to 7 m^2/L (200 to 300 ft^2/gal).

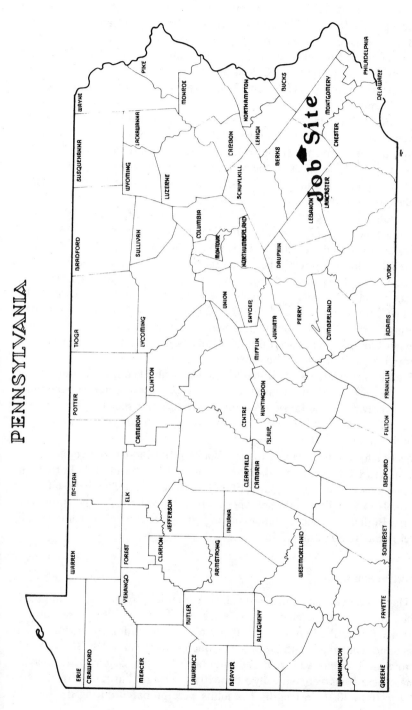

FIG. 1—*Project location map.*

Section 2, PavePrep®—This membrane is produced by McAdams Manufacturing Co., Inc., of Cincinnati, OH. PavePrep membrane consists of a high-density asphaltic mastic between two layers of nonwoven polypropylene fabric for a total membrane thickness of 3 mm (120 mil). This membrane requires the application of AC-20 asphalt cement to bind the fabric to the road surface. PavePrep is 508 mm (20 in.) wide and is cut in 3.7 to 4.6 m (12 to 15 ft) lengths.

Section 3, Bituthene S-5300®—This membrane is manufactured by W. R. Grace and Co., NJ. This membrane is a 1.6-mm (65-mil) thick composite of a polypropylene woven mesh and a layer of self-adhesive rubberized asphalt with a paper release sheet on the bottom. The membrane is 305 mm (12 in.) wide and is supplied in 18.3-m (60-ft) rolls. A Bituthene P-3000 primer is applied at the rate of 6 to 8 m^2/L (250 to 350 ft^2/gal).

Section 4, Roadglas®—This membrane is manufactured by Owens-Corning Fiberglass of Granville, OH. Roadglas consists of a high-strength fiberglass woven fabric mat, which is sandwiched between two layers of hot-poured Roadbond binder. The woven fabric is supplied in 305-mm (12-in.) rolls and cut to the desired length on the job. The polymerized asphaltic binder is supplied prepackaged in plastic bags at a unit weight of 25 kg (55 lb). The binder is heated to 191 to 204°C (375 to 400°F) in a standard heating kettle, poured onto the pavement, and then "squeegeed" into place. The cut fabric is then placed in the hot mastic binder, and another layer of binder is spread on top of the fabric.

Section 5, Petrotac®—This membrane when supplied for this project was called Petromat Y78 and consists of a rubberized asphalt impregnated nonwoven, polypropylene fabric. Petrotac has a release paper backing and is a total of 1.9 mm (75 mil) thick. The membrane is 305 mm (12 in.) wide and is supplied in 30.5-m (100-ft) long rolls.

Section 6, Royston 108 and 10AR—Both the 108 and 10AR membranes are manufactured by Royston Laboratories, Inc., of Pittsburgh, PA. Both membranes consist of an impregnated fiberglass mesh sandwiched between layers of a polymer modified bitumen and a top surface of a woven polypropylene fabric. Both have a release paper backing, are approximately 1.5 mm (60 mil) thick, and are 305 mm (12 in.) wide.

Construction Data

The project consisted of occasional patching, joint preparation, and pavement overlay. The patching was effected in the Fall of 1980 by sawing and removing the full 229 mm (9 in.) depth of broken reinforced concrete pavement and replacing it with bituminous concrete base course (BCBC) material. A total of 37 areas were replaced in either the right, center, or left paving lane. These areas were typically 3.0 m (10 ft) in length but ranged from 1.8 to 12.2 m (6 to 40 ft) in length. The joints were prepared during the week of

13 April 1981 by blowing out the debris, picking out the embrittled PA "J-1," and leveling with PA "FJ-1" bituminous concrete fine aggregate mix. Various heavy duty membranes were then placed over the longitudinal and transverse joints on 16 and 20 April. Next, the pavement was tack-coated with synthetic resin and overlayed with a 50.8-mm (2-in.) layer of PA "ID-2" binder course and a 38.1-mm (1½-in.) layer of PA "ID-2" wearing course. This "ID-2" well-graded bituminous concrete hot mix overlay was placed to the full 88.9 mm (3½ in.) depth on the 9.8-m (32-ft) wide travel lanes. The overlay thickness was feathered down to 25.4 mm (1 in.) at the curb over the new 2.4-m (8-ft) wide shoulders. This overlay was placed on 27 and 28 April.

The heavy duty membranes were placed over transverse, longitudinal, and shoulder/pavement edge joints and over occasional random cracks as detailed in Figs. 2 through 4. Six manufacturers (Table 1) participated by providing materials and installation at no cost to the department while the department provided traffic control, an asphalt heating kettle, and material transportation to the site. Two control sections were constructed without membranes for comparison purposes.

Comments

Table 1 summarizes the material in-place costs and installation rates for each material. This table also includes a listing of ratings for "ease of material application" and for the "effect of traffic" on the in-place membranes. The two membranes, which required the application of a hot bituminous binder, PavePrep and Roadglas, have the highest in-place costs because of the slower application rates and additional manpower and equipment requirements. The installation of these two systems immediately ahead of a paving operation may cause unnecessary delay, and, as such, it is recommended that they be placed a day or two ahead of the scheduled paving operation. These two membranes, along with the Royston 108, which experienced problems with removal of the release paper, were rated low on "ease of application." The cost estimates for these membranes were based upon the time required to place the membranes under research conditions. The actual cost for routine construction projects may vary.

It was not intended that any of these membranes be exposed to direct traffic for more than two or three days; but the first two applied membranes, Roadglas and Petromat Y78 (Petrotac), were exposed eleven days because paving was delayed by poor weather conditions. Both of these membranes were unaffected by traffic, hence the high ratings. The other four membranes were exposed for seven days with varying degrees of deformations.

Performance

Detailed visual inspections were made in Sept. 1981, Aug. 1982, and March 1983. No cracks were evident in any section in Sept. 1981 before the

56   PAVEMENT MAINTENANCE AND REHABILITATION

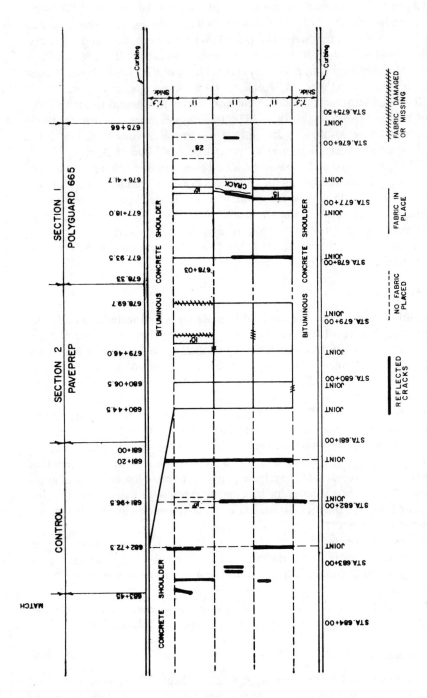

FIG. 2—L.R. 149-22 M. TR 422 (Sections 1, 2, and control).

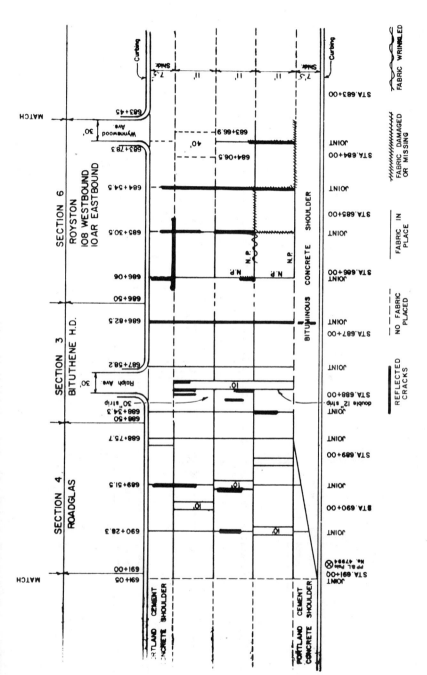

FIG. 3—*L.R. 149-22 M. TR 422 (Sections 3, 4, 8, and 6).*

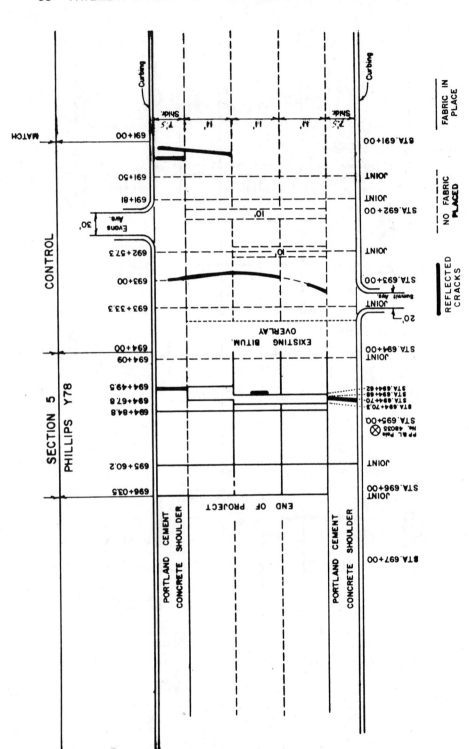

FIG. 4—*I R 140-22 M TR 422 (Section 5 and control).*

TABLE 1—Membrane data summary.[a]

Trade Name	Manufacturer	Material Cost, $/linear ft	In-Place Cost, $/linear ft	Installation Rates	Ease of Application	Effect of Direct Traffic
Polyguard 665® (Section 1)	Polyguard Products, Inc. Pryor, OK	0.39 to 0.42	0.85	P 4 min/100 ft M 8 min/100 ft (2 member crew)	1	2
PavePrep® (Section 2)	McAdams Mfg. Co., Inc. Cincinnati, OH	0.83	1.20	T 37 min/100 ft (3 member crew)	4	3
Bituthene S-5300® (Section 3)	W. R. Grace and Co. North Bergen, NJ	0.34	0.55 to 0.65	P 15 min/100 ft M 15 min/100 ft (2 member crew)	2	2
Roadglas™ (Section 4)	Owens/Corning Fiberglas Granville, OH	fabric 0.22 binder 0.40 0.62	1.27	T 16 min/100 ft (3 member crew) plus 20-min wait	3	1
PetroMat Y78® (Section 5)	Phillips Fiber Corp. Greenville, SC	0.33	0.47	T 10 min/100 ft (2 member crew)	1	1
Royston 108	Royston Lab., Inc. Pittsburgh, PA	0.54	0.78	P 7 min/100 ft M 20 min/100 ft (2 member crew)	3	4
Royston 10AR	Royston Lab., Inc. Pittsburgh, PA	0.53	0.77	M 12 min/100 ft (2 member crew)	2	2

[a] P = application of primer. M = application of membrane. T = combined operation time. Ease of application = No. 1 easiest to apply. Affect of direct traffic = No. 1 least affected. $1/linear ft = 1 ft = 0.3 m.

first freezing season. Summary of the cracks observed in 1982 and 1983 is given in Table 2 and Fig. 5.

The products are ranked (Table 3) from one to seven according to performance, with the No. 1 product having the lowest percentage of reflected cracks.

Table 2 summarizes the reflective crack data as of the March 1983 inspection. Part A of this table shows the total footage of longitudinal and transverse joints in each of the respective pavement sections. Part B lists only that portion of the total joints in each section that remained untreated (no membranes). Note that none of the joints were treated in the two control sections. Part C lists only that portion of the total joints in each section that was treated. Note that all of the joints in Section 4 were treated, and all of the transverse joints in Section 3 were treated.

Three products (PavePrep, Petrotac, and Roadglas) show a substantially lower percentage of reflective cracking over the transverse joints than the other four products. The average amount of cracks reflected over transverse joints treated with these three products is only 6% compared to about 38% reflected over all of the untreated transverse joints after two freezing seasons.

Figure 3 essentially shows the same information as Table 2 but does it in graphical form. This figure does indicate that most of the crack reflection occurred during the first freezing season Aug. 1982 data.

Removal of full-depth pavement cores in Aug. 1983 from treated areas that exhibited reflection cracks revealed that some of the membranes were acting as effective water stops, while other membranes had ruptured. The Polyguard was essentially intact and adhering to both the RCCP and the bituminous concrete binder even though the reflected cracks were as much as 3.2 mm (1/8 in.) wide. The Roadglas and Bituthene had ruptured apparently because the original wide joints in the RCCP were not leveled properly. This resulted in the membranes undergoing abrupt, differential, vertical deflections and subsequent rupture from shear. These membranes were no longer functional as a water stop. Proper preparation of the old joints before membrane application would tend to minimize this failure mode. Claims by some membrane producers that these products can span totally unprepared joints are questionable.

Cost Effectiveness

The cost effectiveness and service life of these membranes have not been determined. It is believed that membranes may delay the development of reflection cracks but not prevent their eventual formation. The value of a two or three year delay in the formation of a crack in a pavement with a service life of ten years is questionable. The cost of sealing the crack an additional time or two would be small. The real value of these membranes may be in waterproofing the joint.

TABLE 2—*Summary of field inspection, to March 1983.*[a]

Section	A Total of Joints and Cracks, ft		A Total, ft		B. Untreated Joints and Cracks						C. Treated Joints and Cracks			
					Total of Reflected Cracks, ft		Reflected Cracks, %		Total, ft		Reflected Cracks, ft		Reflected Cracks, %	
	Trans.	Longi-tudinal	Trans.	Longi-tudinal	Trans.	Longi-tudinal	Trans.	Longi-tudinal	Trans.	Longi-tudinal	Trans.	Longi-tudinal	Trans.	Longi-tudinal
1, Polyguard 665	220	1067	44	287	17	0	38.6	0	176	780	37	0	21.0	0
2, Pave-Prep	154	668	22	143	0	0	0	0	132	525	0	0	0	0
Control 1	135	980	135	980	114	0	84.4	0	...	...	...	...	...	...
6, Royston 108[b]	66	610	46	306	29	0	63.0	0	20	304	8	0	40.0	0
10AR	110	610	55	535	10	98	18.2	18.3	55	75	24	0	43.6	0
3, Bituthene H.D.	186	800	0	476	0	0	0	0	186	324	71	0	38.2	0
4, Roadglas	196	1000	0	0	0	0	0	0	196	1000	21	0	10.7	0
Control 2	337	1200	337	1200	66	0	19.6	0	...	...	...	...	...	...
5, Phillips Petrotac	215	816	15	224	12	0	80.0	0	200	592	10	0	5.0	0
Total	1619	7751	654	4151	248	98	37.9	2.4	965	3600	171	0	17.7	0

[a] 1 ft = 0.30 m.
[b] Damaged areas counted as untreated.

62 PAVEMENT MAINTENANCE AND REHABILITATION

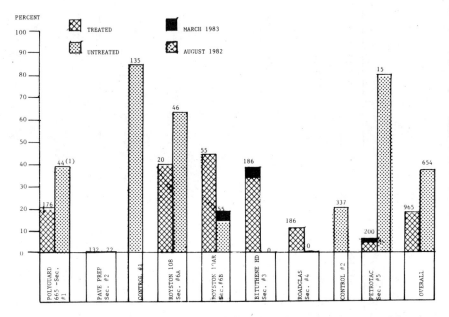

FIG. 5—*Transverse reflective cracking. (1) Indicates number of feet of transverse joints that were treated or untreated upon which the percentage of reflective cracking is based or calculated. Applies to all column numbers. 1 ft = 0.30 m.*

TABLE 3—*Membrane performance.*

Product	Treated Transverse Cracks, % Reflected	Ranking
Pave-Prep	0.0	1
Petrotac	5.0	2
Roadglas	10.7	3
Polyguard 665	21.0	4
Bituthene H.D.	38.2	5
Royston 108	40.0	6
Royston 10AR	43.6	7

Alternates to the placement of membranes and their costs are as follows:

1. Prepare the joint, reseal, and resurface, and when the reflection crack forms, seal with crack sealer. This procedure is the one most commonly used. The cost to seal cracks is approximately $0.19/m ($0.06/linear ft). This procedure has the lowest initial cost, and while it is satisfactory for pavements of good design and proper compaction, it presents many problems for most overlays. Some of these problems are as follows: (1) When there are random cracks in the RCC pavement, the multiple reflection cracks may lead to block

cracking, (2) raveling may occur along the cracks and permit water to enter the subbase, and (3) the reflection cracks present a continuing maintenance problem.

2. Prepare the joint, reseal, and resurface. Then saw a transverse joint in the bituminous concrete overlay, and seal the sawed joint. The cost of this procedure is approximately $1.67/m ($0.50/linear ft). This procedure is being used more frequently and is suitable when the RCCP is structurally sound. It is not satisfactory for pavements that have deteriorated joints or multiple cracks. The advantage of this procedure is that the sawed joint is smooth and of the proper shape for sealing, and it helps prevent random cracking, raveling, and secondary shear cracking.

3. Remove and replace the joint. This procedure is the most costly, approximately $21.30/m ($6.39/linear ft). But when the RCCP has a good riding surface and has few cracks, this procedure may rehabilitate the roadway without resurfacing. This procedure introduces two potential points (at both ends of the patch) for reflection cracks for each joint repaired when the pavement is overlayed.

The decision to use membranes must be made on a job by job basis, after evaluating the condition of the RCCP and other available alternates.

Summary

The following conclusions are drawn:

1. There was significant differences in the total length of reflected cracks in the two control sections, 84% in Control 1 and 20% in Control 2. This difference, between control sections, along with the large difference among the control sections and the untreated areas in the adjacent test sections, made it difficult to use the control sections for comparison with treated areas. Therefore, the performances of all products were compared to that of the total project untreated joints and cracks.

2. There was a reduction of about 20% in the reflection cracks in the total treated areas versus total untreated areas, 38% of the untreated area reflected versus 18% of the treated area. The membranes, as a whole, are retarding reflective cracking.

3. Three products (PavePrep, Petrotac, and Roadglas) are performing significantly better than the other four after two freezing seasons. The average amount of cracks reflected over transverse cracks treated with these three products is only 6% compared to about 38% reflected over all of the untreated transverse joints.

4. Approximately 98% of the reflection cracking over transverse joints and cracks occurred during the first freezing season. No treated longitudinal cracks had reflected after two seasons compared to only about 2% crack reflection over untreated longitudinal joints.

5. Some products continued to act as an effective water stop even after the reflection crack had propagated to the new surface and opened. Other products ultimately ruptured from shearing forces and failed to provide an effective water stop.

6. The length of treated joints and cracks in the Royston Section, 6 m (20 ft) in Section 6A and 17 m (55 ft) in Section 6B, may be too small for a valid test.

7. The cost/benefits of using membranes has not been determined. This evaluation will be continued to determine whether the cracking has leveled off or will continue to form at various rates, depending on temperature extremes in future winters. Additional cores will be removed from this project to determine the condition of the membranes in those areas where reflection cracks have not developed. The positive effects that the water stopping properties of the membranes may have on future pavement joint deterioration rates will also be monitored.

Reference

[1] Wouter Gulden, P. E., "Rehabilitation of Plain Portland Cement Concrete Pavement with Asphaltic Concrete Overlays," Georgia Department of Transportation, Office of Materials and Research, Forest Park, GA, 1978.

R. David Rowlett[1] and William E. Uffner[1]

The Use of an Asphalt Polymer/Glass Fiber Reinforcement System for Minimizing Reflection Cracks in Overlays and Reducing Excavation Before Overlaying

REFERENCE: Rowlett, R. D. and Uffner, W. E., "**The Use of an Asphalt Polymer/Glass Fiber Reinforcement System for Minimizing Reflection Cracks in Overlays and Reducing Excavation Before Overlaying,**" *Pavement Maintenance and Rehabilitation, ASTM STP 881*, B. F. Kallas, Ed., American Society for Testing and Materials, Philadelphia, 1985, pp. 65-73.

ABSTRACT: Many techniques have been used in an attempt to prevent reflection cracking in asphalt overlays including stress relieving membranes, fabrics, bond breakers, and changes in mix design. Glass fiber reinforcements have been used in the plastics industry for increasing the fatigue life and fracture toughness of plastic products for many years. This paper describes the successful development of the use of glass fiber reinforcement in conjunction with an asphalt polymer to minimize reflection cracks in an asphalt overlay. The system is also used to reduce excavation of deteriorated cracks and joints before overlaying.

The glass fiber reinforcement is a high strength woven glass with a tensile strength of 1750 N/cm (1000 lb/in.) width. The asphalt polymer was specially designed for use with the glass fiber reinforcement and has excellent properties of adhesion, low-temperature flexibility, and high softening point.

The combined properties of the system are the key to performance. The asphalt polymer adheres to the old road, the glass fiber reinforcement, and the new overlay, protects the glass fiber reinforcement, acts as a stress relieving membrane, and transfers the stresses to the reinforcement. The glass fiber reinforcement distributes the stress over the width of the repair to a level below the fracture point of the overlay. Laboratory tests were used to determine the physical properties and characteristics of the system, but no laboratory tests were found to predict the field performance. Evaluation of the system was performed through field trials. Over a five-year period 150 field trials were installed in 21 states. These trials included transverse and longitudinal cracks in bituminous roads and transverse joints, longitudinal joints, and interslab cracks in portland cement concrete base roads. Approximately 30 480 m (100 000 ft) of the repairs were monitored for crack

[1] Laboratory supervisor and research associate, respectively, Research and Development Division, Owens-Corning Fiberglas Corporation, Technical Center, Granville, OH 43023.

reflection. The asphalt polymer/glass fiber reinforcement system was 89% effective in reducing reflection cracks. The average age of the trials was 21 months with a range of 6 to 56 months. All trials were exposed to a minimum of one winter.

In recent trials this system has been used to reinforce badly distressed areas around joints or cracks in lieu of partial or full depth excavation. Cost savings greater than ten to one have been demonstrated for the asphalt polymer/glass fiber reinforcement system with equivalent or better performance than excavation.

KEY WORDS: reflection, cracks, glass fibers, covering, excavation, tensile strength, reflection crack, glass fiber reinforcement, asphalt polymer, low temperature flexibility

Reflection cracks have been identified as a major contributor to the deterioration of overlayed roads. Many techniques have been used in an attempt to prevent reflection cracking in asphalt overlays including stress relieving membranes, fabrics, bond breakers, and changes in mix design with limited success. Glass fiber reinforcements have been used in the plastics industry for increasing the fatigue life and fracture toughness of plastic products for many years. The combination of two technologies, glass fiber reinforcement and chemical modification of asphalt, led to the successful development of the use of glass fiber reinforcement in conjunction with an asphalt polymer to minimize reflection cracks in an asphalt overlay. The system is also used to replace excavation of deteriorated cracks and joints before overlaying.

Experimental Approach

Through discussions with highway engineers, consultants, paving contractors, and others knowledgeable in the field, it was determined that no laboratory tests could predict the field performance of systems designed to reduce reflective cracks. Early field trials were used to screen many types of fabrics and asphalts. These trials indicated the need for an asphalt polymer specifically designed to survive construction traffic and paving, and still contribute to reduced reflective cracking. In addition, these early trials indicated that the system of a high strength woven glass fiber reinforcement and an asphalt polymer had the best potential for reducing reflection cracks. Laboratory tests were used to determine the physical properties and characteristics of the system and its components, and field trials were used to evaluate the performance of the system.

Description of the System

The system developed for reduction of reflection cracks is a combination of a high tensile strength, high modulus woven glass fiber fabric, and an asphalt polymer with properties designed for use with the glass fabric. Adhesion, low-temperature flexibility, and high softening point properties provide the key to performance. The asphalt polymer adheres to the old road, the glass fiber reinforcement and the new overlay, protects the glass fiber reinforcement, acts as a stress relieving membrane, and transfers the stresses to the reinforcement. The glass fiber reinforcement distributes the stresses over the width of the repair to a level below the fracture point of the overlay.

This system is a spot repair system and each crack or joint is repaired. The installation procedure is

1. Clean out the cracks with compressed air or broom. The road surface and crack must be dry.
2. Melt the asphalt polymer (usually in a tar kettle) and heat to 464 K (375°F). Care must be taken not to heat the polymer over 477 K (400°F) or degradation can occur.
3. Cut fabric to the length of crack.
4. Pour molten asphalt polymer over crack.
5. Spread with a squeegee to form a smooth, continuous, and uniform surface.
6. Place the glass fiber reinforcement over the crack while the asphalt polymer is still molten.
7. Pour molten asphalt polymer and squeegee, a top coating, over the glass fiber reinforcement.

The repair cools in approximately 30 min and may be exposed to traffic for a short time or paved over immediately. The application rate is approximately 0.004 m^3/m^2 (0.1 gal/ft^2), and the reinforcement is available in 0.3, 0.6, and 1.1 m (12, 24, and 44 in.) widths. The rheological and adhesive properties of the asphalt polymer allow the repair to be paved over without pull-up, tracking, or bleeding into the new overlay.

Laboratory Results and Discussion

Laboratory testing was used to characterize the physical properties of the asphalt polymer, glass fiber reinforcement, and the coated reinforcement, and to gain insight into the mechanism by which reflection cracks were suppressed. Typical properties of the asphalt polymer and the glass fiber reinforcement are given in Tables 1 and 2, respectively. Some properties of a typical AC-20 (viscosity graded asphalt cement) and a nonwoven polypropylene fabric are given for comparison.

Inspection of these physical properties indicates some of the unique properties this system exhibits that contribute to its performance during installation and paving and in suppressing reflection cracks.

The asphalt polymer exhibits good properties over a wide temperature range as indicated by the softening point, hot viscosity, low temperature flexibility, and elastic modulus. The high softening point and viscosity at paving temperatures allows the asphalt polymer to survive the paving operation without tracking under traffic or bleeding through the new pavement. However,

TABLE 1—*Properties of the asphalt polymer used in the asphalt polymer/glass fiber reinforcement system for suppression of reflection cracks.*

Property	Test Temperature, K (°F)	Test Method	Asphalt Polymer	AC-20
Softening point, K (°F)	...	ASTM Test Method for Softening Point of Bitumen (Ring-and-Ball Apparatus) (D 36)	355 (180)	325 (126)
Penetration, 0.1 mm	298 (77)	ASTM Test Method for Penetration of Bituminous Materials (D 5)	54	52
Flash point, K (°F)	...	ASTM Test Method for Flash and Fire Points by Cleveland Open Cup (D 92)	563 (585)	543 (550)
Lap shear adhesion, kPa (psi)	298 (77)	ASTM Recommended Practice for Determining the Strength of Adhesively Bonded Plastic Lap Shear Sandwich Joints in Shear by Tension Loading (D 3164)	284 (36)	79 (11)
Peel adhesion,[a] kPa (psi)	298 (77)	ASTM Test Methods for Peel Adhesion of Pressure Sensitive Tape at 180° Angle (D 3300)	55 (8)	17 (2.5)
Low temperature flexibility,[b] K (°F)	...	ASTM Recommended Practice for Flexibility Determination of Hot Melt Adhesives by Mandrel Bend Test Method (D 3111)	250 (−10)	273 (32)
Hot viscosity, Pa·s (cP)	466 (380)	ASTM Test Method for Apparent Viscosity of Hot Melt Adhesives and Coating Materials (D 3236)	1.2 (1 200)	0.04 (42)
	411 (280)	...	13.0 (13 000)	0.30 (300)
Solution viscosity, Pa·s (cP)	298 (77)	Seventy percent solids in toluene using brookfield RVT viscometer	5.2 (5 200)	2.0 (2000)
Permeance to water, ng/Pa·s·m² (perms)	...	ASTM Test Methods for Water Vapor Transmission of Materials (E 96)	2.9 (.05)	2.9 (.05)
Elastic modulus, MPa (psi)	277 (40)	ASTM Test for Tensile Properties of Plastic (D 638)	6.2 (896)	35.7 (5 180)
...	261 (10)	...	20.6 (2 981)	270.0 (39 200)
...	250 (−10)	...	62.5 (9 050)	...[c]
...	233 (−40)	...	396.0 (57 380)	...[c]

[a] Tape made from glass fiber reinforcement and the bond was formed with molten asphalt polymer.
[b] Mandrel diameter 2.54 cm (1 in.) and specimen size 3.8 by 22.9 by 0.38 cm (1.5 by 9 by 15 in.).
[c] Not able to test material, shattered in grips.

TABLE 2—*Properties of glass fiber reinforcement for suppression of reflection cracks.*

Property	Test Method	Glass Reinforcement	Nonwoven Polypropylene Mat
Weight, g/m² (oz/yd²)	ASTM Test for Weight (Mass) per Unit Area of Woven Fabric (D 3776) and ASTM Testing Nonwoven Fabrics (D 1117)	568 (24)	95 (4)
Tensile strength, N/cm (lb/in.) width	ASTM Specification for Greige Woven Glass Fabrics (D 579) Procedure 2 and ASTM Test for Nonwoven Fabrics (D 1117)	1750 (1000)	158 (90)
Initial modulus,[a] MPa (psi)	calculated from grab tensile force-elongation curves at break	230 (33 500)	6.2 (900)

[a] Across the crack.

the viscosity at application temperature is low enough to allow easy installation. In contrast AC-20 has a softening point lower than road surface temperatures on a hot sunny day (approximately 344 K [160°F]), which can lead to tracking, and has viscosity characteristics that make it susceptible to bleeding during paving.

The asphalt polymer also has excellent low-temperature properties as indicated by a low-temperature flexibility of 250 K (−10°F) and an elastic modulus of 62.4 MPa (9000 psi) at 250 K (−10°F) to 396 MPa (57 000 psi) at 230 K (−40°F). A typical bituminous hot mix, such as the Ohio Department of Transportation 404 mix, has an elastic modulus in the range of 19.0 to 26.9 GPa (2.75 to 3.9 million psi) for a temperature range of 250 to 233 K (−10 to −40°F). By comparison, an AC-20 had a low temperature flexibility of 273 K (32°F), and the elastic modulus could not be measured at 250 K (−10°F) because the specimen shattered.

The low-temperature properties are significant in the performance of the asphalt polymer/glass fiber reinforcement system in suppressing reflection cracks. The elastic properties of the asphalt polymer at low temperatures, when the thermal stresses are greatest, enable the system to absorb the imposed stresses and transfer much of the stress to the higher strength glass reinforcement. A brittle coating material would not allow the stress absorption and transfer. The good adhesive properties are also important for (1) preventing pull-up during installation and paving and (2) providing a good bond between the old and new pavement and the glass reinforcement to aid in stress transfer. Other properties of the asphalt polymer to note include the solution viscosity, elastic modulus, and hot viscosity, which indicate its polymeric nature. The water vapor transmission rate indicates its ability to waterproof the repair area.

The significant properties of the glass fiber reinforcement are the tensile strength and initial modulus. The combination of high tensile strength and high modulus allow the distribution of stresses across the width of the glass but prevents the strain from being transferred to the new overlay. The combination of the glass fiber reinforcement and the asphalt polymer increases the tensile strength of the system to 2100 N/cm (1200 lb/in.) width from 1750 N/cm (1000 lb/in.) width for uncoated glass fiber reinforcement. Many asphalts and rubberized asphalts will decrease the tensile strength of the coated system.

The asphalt polymer/glass fiber reinforcement system was tested in a load fatigue simulation by Resource International, Inc. [1]. The test consists of a 7.6- by 7.6- by 61-cm (3- by 3- by 24-in.) asphalt concrete beam. The test material is placed on the first inch of the beam and then covered with the next two inches of asphalt concrete. The bottom 7.6-cm (1-in.) layer is notched to simulate a crack. The test beam is placed on an elastic foundation chosen to simulate a typical road subbase. A 90.7-kg (200-lb) load is cycled over the notch at 2 Hz. The test is run at 298 K (77°F) until the crack propagates to the top of the test beam. The results of this testing indicate an increase in load fatigue life of the asphalt concrete when the asphalt polymer/glass fiber reinforcement system is used to repair the crack. The asphalt polymer/glass fiber reinforcement system did not fail after 327 000 cycles (the test was stopped before complete failure), but the control failed after 7300 cycles.

Field Trial Results and Discussion

The proof of a system for reducing reflection cracks is in field testing. Over a five-year period 150 field trials were installed in 21 states to determine the performance and commercial viability of the asphalt polymer/glass fiber reinforcement system. These trials included transverse and longitudinal cracks in bituminous roads and transverse joints, longitudinal joints, and interslab cracks in portland cement concrete (PCC) base roads. These trials were installed with available labor under Owens-Corning Fiberglas supervision. Approximately 30 480 m (100 000 ft) of the repairs were monitored for crack reflection. A summary of the results are given in Table 3. Overall the asphalt polymer/glass fiber reinforcement system was 89% effective in reducing reflection cracks over the time period studied. The average age of all the trials was 21 months with a range of 6 to 56 months. All trials were exposed to a minimum of one winter. The effectiveness is calculated as the linear feet of cracks repaired that do not reflect into the new overlay as a percent of the linear feet repaired. The results are divided into four categories to emphasize the variability between the different types of cracks and joints that can reflect through a new overlay. In particular, concrete transverse joints are the most difficult to repair effectively because of vertical deflection, voids under the slab ends, and in many cases severe deterioration of the concrete in the area

TABLE 3—Field trial performance summary of asphalt polymer/glass fiber reinforcement system for reduction of reflection cracks.

Repair Method	Percent Reflected		Length Repaired, m (ft)	Number of Trials	Number of States	Time Frame	Average Age, months	Range[a] of Ages, months
	Average	Range						
			BITUMINOUS CONCRETE					
Asphalt polymer/glass fiber reinforcement system	12	0 to 17.5	20 000 (65 638)	13	6	1976 to 1980	24	10 to 56
Control	88	0 to 100						
			CONCRETE INTERSLAB CRACKS					
Asphalt polymer/glass fiber reinforcement system	1	0 to 1	4 295 (14 092)	3	1	1978 to 1980	19	6 to 32
Control	75	48 to 100						
			CONCRETE LONGITUDINAL JOINTS[b]					
Asphalt polymer/glass fiber reinforcement system	0	0	2 794 (9 167)	11	6	1978 to 1980	15	6 to 30
Control	5	0 to 15						
			CONCRETE TRANSVERSE JOINTS					
Asphalt polymer/glass fiber reinforcement system	29	0 to 75	2 244 (7 362)	13	6	1978 to 1980	17	6 to 30
Control	86	0 to 100						

[a] All trials were exposed to a minimum of one winter.
[b] Including widening seams.

around the joint. This currently unpredictable situation is reflected in the wider range of performance on PCC joints.

In other trials the asphalt polymer/glass fiber reinforcement system has been used to reinforce badly distressed areas around joints or cracks in lieu of partial or full depth excavation.

The steps in a full depth excavation on a composite road are:

(1) identify the distressed area,
(2) saw out, break up, and remove overlays and base pavement, and
(3) replace with PCC or flexible pavement or both.

For a partial depth excavation the steps are:

(1) identify the distressed area,
(2) saw out, break up, and remove the overlays (leave base pavement intact), and
(3) replace with flexible pavement.

The objective of this application is to greatly reduce the cost of repair while maintaining equivalent or better overlay performance. An example of the application is on State Route 16 near Pataskala, OH. A summary of this trial and a comparison of actual costs for repair are given in Table 4. The average width of the repairs was 1.2 m (4 ft). Evaluation of the road after three winters using a pavement condition rating system, jointly developed by Owens-Corning Fiberglas and the Ohio Department of Transportation, indicates equivalent performance of the two systems. Evaluation using ridability and roughness indicates that the asphalt polymer/glass fiber reinforcement system has performed better than excavation. This evaluation is a subjective assessment made while driving over the trial site.

Conclusions

Based on extensive field trial data, it was concluded that the asphalt polymer/glass fiber reinforcement system developed by Owens-Corning Fiberglas

TABLE 4—*Asphalt polymer/glass fiber reinforcement system as for excavation on State Route 16, in Pataskala, OH, average daily traffic 8930, trucks 8%.*

Repair Method	Overlay Thickness, cm (in.)	Length of Joints Repaired, m (ft)	Number of Joints Repaired	Per Joint Actual Cost, $
Asphalt polymer/glass fiber Reinforcement system	2.54 (1)	238 (780)	65	58
Control excavated (adjacent)[a]	2.54 (1)	238 (780)	65	811

[a] 46% of the excavations were partial depth at $550 per joint and 55% of the excavations were full depth at $1033 per joint.

is effective in reducing reflection cracks in asphalt overlays. In addition, recent trials have demonstrated first cost savings greater than ten to one for repair of badly distressed areas around joints or cracks in lieu of partial or full depth excavation.

This product was commercialized in 1980 based on the data presented in this paper [2].

References

[1] Majidzadeh, K., Dat, M., and Makdisi-Ilyas, F. "Application of Fracture Mechanics for Improved Design of Bituminous Concrete. Volume 2, Evaluation of Improved Mixture Formulations and the Effect of Temperature Conditions on Fatigue Models," FHWA-RD-76-92, Ohio State Research Foundation, Department of Civil Engineering, Columbus, OH, 1976.

[2] "ROADGLAS®, The Engineered Solution to Reflective Cracking," Owens-Corning Fiberglas Publication 5-R5-11489, Owens-Corning Fiberglas Corporation, Toledo, OH, 1982.

Mang Tia[1] *and Leonard E. Wood*[2]

The Use of a Gyratory Testing Machine in the Evaluation of Cold-Recycled Asphalt Paving Mixtures

REFERENCE: Tia, M. and Wood, L. E., **"The Use of a Gyratory Testing Machine in the Evaluation of Cold-Recycled Asphalt Paving Mixtures,"** *Pavement Maintenance and Rehabilitation, ASTM STP 881*, B. F. Kallas, Ed., American Society for Testing and Materials, Philadelphia, 1985, pp. 74–90.

ABSTRACT: This study evaluated the feasibility of using the gyratory testing machine in the fixed roller mode for evaluating the long-term behavior of cold-recycled asphalt mixtures. Cold recycled asphalt mixtures were compacted with the gyratory machine, and gyratory strain indices were obtained from the gyrograph recorded during the compaction process. The resilient modulus, Hveem stabilometer R value, and Marshall stability were obtained on the compacted recycled mixes at various curing times and were correlated to the gyratory strain indices. The scope of study presented in this paper covered two types of pavement material and five types of added softening agents, which included a high-float asphalt emulsion AE-150, a foamed asphalt and the rejuvenating agents, Reclamite®, Mobilsol®, and Dutrex 739®.

The results of this study indicated that the gyratory stability index (GSI) and the gyratory elasto-plastic index (GEPI) could be used to detect unstable mixtures when the binder content was too high. The GSI and GEPI correlated fairly well with the Hveem stabilometer R value of the recycled mix. However, they correlated poorly with the resilient modulus and Marshall stability of the recycled mix. Similar to the R-value test, the gyratory testing machine can be used to determine the optimum binder content of a cold-recycled mix.

The scope of this study was limited to the use of the gyratory testing machine in the fixed roller mode. Tests using the oil filled roller mode had not been conducted in this investigation.

KEY WORDS: flexible pavements, pavements, asphalt emulsions, cold-recycled asphalt mixture, gyratory testing machine, gyratory index, resilient modulus, R value, Marshall stability, foamed asphalt, rejuvenating agents

[1]Assistant professor, Department of Civil Engineering, Weil Hall, University of Florida, Gainesville, FL 32611.
[2]Professor, School of Civil Engineering, Civil Engineering Building, Purdue University, West Lafayette, IN 47907.

Asphalt pavement recycling has become an important pavement rehabilitation method in recent years. The increased importance of asphalt pavement recycling has made it necessary to understand the behavior of the recycled asphalt mixtures more fully and to examine the testing methods that may be suitable to be used in the evaluation of these mixes. Basically, asphalt pavement recycling involves the addition of virgin aggregates to upgrade the old aggregates and the addition of rejuvenating agents and new asphalts to soften the hardened old asphaltic binders. In hot-mix recycling, the blending of the old asphaltic binders, and the new asphalts or rejuvenating agents is more homogeneous, and the behavior of the hot-recycled mixtures is similar to that of a conventional asphalt mix. Cold-mix recycling is slightly different. During the cold recycling process, a thin film of virgin binder or softening agent is introduced. This thin film of virgin binder or softening agent will have a rejuvenating effect on the old binder material. The rejuvenating action that takes place may be dependent on time, temperature, and additional traffic compaction. In the design of cold-recycled asphalt mix, it is desirable to be able to predict the long-term behavior from the short-term laboratory results.

This study investigated the feasibility of using the gyratory testing machine in the fixed roller mode for evaluating the long-term behavior of cold recycled asphalt mixtures. In a previous study on cold-mix recycling by the authors [1], it was found the potential problem of instability of a recycled mixture, which may show up some time after construction, could be detected when the mixture was subjected to a high compactive effort with the gyratory compactor. It was thought that the compactive effort of the gyratory machine forced the old binder and the virgin binder or softening agent to act as one. Thus, the rejuvenating action was expedited. Based on this hypothesis, the gyratory machine operated in the fixed roller mode was used to compact cold-recycled asphalt mixtures and to evaluate their performance. The study has the following objectives:

(1) to evaluate the feasibility of using the gyratory testing machine operated in the fixed roller mode to determine the optimum binder contents of cold-recycled asphalt mixtures and

(2) to evaluate the correlation of the gyratory strain indices obtained during the gyratory compaction process with the resilient modulus, Marshall stability, and Hveem R values of the compacted cold-recycled asphalt mixtures.

Equipment and Material

Equipment

The major pieces of equipment used in this study include the gyratory testing machine, the resilient modulus test equipment, the Hveem stabilometer and compression machine, the Marshall testing equipment, and the Foamix asphalt dispenser.

The gyratory testing machine in the fixed roller mode was used for compaction and testing of the recycled mixtures. Figure 1 shows the cross section through the gyratory mechanism. A specimen in a mold is held by a mold chuck and a constant ram pressure. The mold clamped in the mold chuck is set at a fixed angle (initial angle of gyration) by two fixed rollers. The bottom roller is adjustable but is held fixed at the selected angle. As the two fixed rollers rotate, a shear strain is constantly applied to the mixture in the mold, and as a result, the mixture is compacted by a gyratory kneading mechanism. A ram pressure of 1.38 MPa (200 psi) and an initial gyratory angle of 1° were used. The gyragraph recorder plots the shear displacement (gyratory angle) of the specimen during the compaction. A recording of the gyratory angle is known as the gyrograph and is used to obtain the gyratory indices.

The diametral resilient modulus test proposed by Schmidt [2] was modified and used in this study. During the resilient modulus test, a 222-N (50-lb) pulse load of 0.1 s duration is applied diametrically to the test specimen every 3 s. The vertical and horizontal deformations of the specimen were measured and used to calculate the resilient modulus of the compacted recycled mixture.

A Hveem stabilometer and a compression machine, which met the ASTM standards, were used for the Hveem R value test in this study.

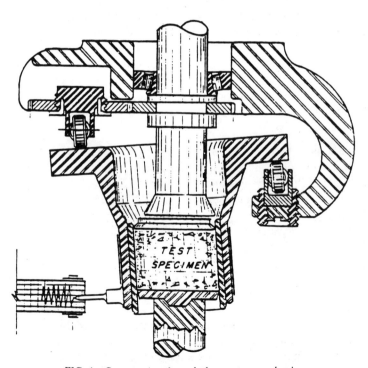

FIG. 1—*Cross section through the gyratory mechanism.*

An autographic Marshall testing apparatus was used to conduct the Marshall stability tests on the recycled mixtures.

A laboratory Foamix asphalt dispenser developed by CONOCO Inc. was used to produce the foamed asphalt to be added to the recycled mixtures.

Material

Pavement Materials—Two pavement materials were used to make the cold-recycled asphalt mixtures in this study. The two pavement materials were an old pavement material and an artificially aged paving mixture.

The old pavement material was obtained from a state road in Indiana. The material had been crushed to a maximum size of 2.54 cm (1 in.). A centrifuge extraction of bitumen (ASTM Test Method for Quantitative Extraction of Bitumen from Bituminous Paving Mixtures [D 2172] Method A) performed on the pavement material indicated an average bitumen content of 5.0% by weight of the aggregate. Recovered bitumen using the Abson recovery method (ASTM Test Method for Recovery of Asphalt from Solution by Abson Method [D 1856]) had a penetration of 25 at 24°C and an absolute viscosity of 6380 Pa·s (63 800 poise) at 60°C. The recovered aggregate consisted mainly of crushed limestone. As compared to Indiana's Type II No. 9 surface mix aggregate, the recovered aggregate was slightly lacking in the coarser sizes.

The artificially aged mixture was made to resemble the old pavement material described earlier. The aggregate used was a limestone, and the gradation was the same as that of the old aggregate. The aggregate was mixed with 5.0% of AP-3 grade asphalt cement by weight of the aggregate at 150°C. The mixture was then artificially aged by placing it in a forced draft oven at 120°C for 24 h. The recovered asphalt from the artificially aged mixture had a penetration of 24 at 25°C and an absolute viscosity of 3230 Pa·s (32 300 poise) at 60°C.

Softening Agents—Softening agents used in this study included (1) a highfloat anionic asphalt emulsion, designated as AE-150 in the Indiana State Highway standard specification, (2) a foamed asphalt made from a soft asphalt designated as AC-2.5, and (3) three rejuvenating agents, Reclamite®, Mobilsol®, and Dutrex 739®. Reclamite and Mobilsol were in the forms of emulsion, while Dutrex 739 was a highly viscous oil. Physical properties of AE-150, AC-2.5, Reclamite, Mobilsol, and Dutrex 739 are described in Tables 1 through 4.

Design of the Experiment

The main objective of this study was to evaluate the feasibility of using the gyratory testing machine to predict the long-term performance of cold-recycled asphalt paving mixtures. Investigation was to be made to determine how well the gyratory strain indices (obtained during the gyratory compaction pro-

TABLE 1—*Physical properties of AE-150.*

Property	Standard	Test Condition	Value
Percent residue by distillation	ASTM D 244[a]	standard	70.0%
Oil portion of distillate	ASTM D 244	standard	1.5%
Tests on distillation residue			
Penetration	ASTM D 5[b]	100 g, 5 s, 25°C	215 (dmm)
Specific gravity	ASTM D 70[c]	25°C	1.010
Float	ASTM D 139[d]	60°C	200 s

[a] ASTM Testing Emulsified Asphalts (including Tentative Revision) (D 244).
[b] ASTM Test Method for Penetration of Bituminous Materials (D 5).
[c] ASTM Test Method for Specific Gravity of Semi-Solid Bituminous Materials (D 70).
[d] ASTM Float Test for Bituminous Materials (D 139).

TABLE 2—*Physical properties of AC-2.5.*

Property	Standard	Test Condition	Value
Penetration	ASTM D 5	100 g, 5 s, 25°C	>300 (dmm)
Absolute viscosity	ASTM D 2171[a]	60°C	30 Pa·s (300 poise)
Kinematic viscosity	ASTM D 2170[b]	135°C	160 cSt
Specific gravity	ASTM D 70	25°C	1.024
Ductility	ASTM D 113[c]	25°C	>100 cm

[a] ASTM Test for Viscosity of Asphalts by Vacuum Capillary Viscometer (D 2171).
[b] ASTM Test Method for Kinematic Viscosity of Asphalts (Bitumens) (D 2170).
[c] ASTM Test Method for Ductility of Bituminous Materials (D 113).

TABLE 3—*Properties of Reclamite and Mobilsol.*

Property	Standard	Test Condition	Values Reclamite	Values Mobilsol
Percent residue	evaporation	heating up to 120°C with a direct flame	63%	72%
Specific gravity of residue	ASTM D 70	25°C	0.966	0.974
Saybolt furol viscosity	ASTM D 88[a]	25°C	25 s	69 s
Molecular analysis	...	...	...	...
asplatenes	...	...	0.5%	...
polar compounds	...	...	17.7%	17.1%
1st acidaffins	...	...	16.4%	16.0%
2nd acidaffins	...	...	38.6%	39.5%
saturates	...	...	26.8%	27.5%

[a] ASTM Test Method for Saybolt Viscosity (D 88).

cess) correlated with the long-term mixture properties. The mixture properties to be used in the correlation study included the resilient modulus, Hveem stabilometer R value, and Marshall stability. The following sections describe the gyratory strain indices, the mixture properties to be correlated, and the experimental design for this study.

TABLE 4—*Properties of Dutrex 739.*

Property[a]	Value
Saybolt furol viscosity (at 25°C), s	9600 s
Specific gravity (at 16°C)	1.0344
Distillation test, °C (°F)	
initial boiling point	393 (740)
5%	407 (765)
10%	411 (772)
50%	437 (818)
90%	473 (884)
Molecular analysis, %	
asphaltenes	0
polar compounds	18.9
aromatics	75.1
saturates	6.0
Estimated molecular weight	340

[a] All the information with the exception of Saybolt furol viscosity, were supplied by Shell Development Company.

Gyratory Strain Indices

As a specimen is being compacted in the gyratory machine, the gyratory motion experienced by the specimen is recorded by a gyrograph, and the magnitude of the gyratory angle is indicated by the width of the gyrograph. Gyratory Indices can then be obtained from the gyrograph.

Gyratory Elasto-Plastic Index

The gyratory elasto-plastic index (GEPI) is defined as

$$\text{GEPI} = \frac{\text{minimum intermediate gyrograph width}}{\text{initial gyratory angle}}$$

It gives an indication of the amount of shear strain (elastic or plastic or both) under an induced amount of fixed initial shear strain. It also indicates the lubricating effect of the bitumen at each increment of increase in bitumen content.

Gyratory Stability Index

The gyratory stability index (GSI) is defined as

$$\text{GSI} = \frac{\text{maximum gyrograph width}}{\text{minimum intermediate gyrograph width}}$$

It gives an indication of the stability of the mixture under compaction. An increase in the stability index would indicate a reduction in stability or shear resistance.

Gyratory Compactibility Index

The gyratory compactibility index (GCI) is defined as

$$\text{GCI}_x = \frac{\text{unit weight at } x \text{ revolutions}}{\text{unit weight at 60 revolutions}}$$

It gives an indication of the relative degree of compaction at a specific compactive effort.

Mixtures Properties Measured

Resilient Modulus M_R—The resilient modulus is defined as the ratio of the applied stress to the resilient strain (recoverable strain) when a dynamic load is applied. It is the dynamic elastic modulus of a viscoelastic material.

Stabilometer Resistance Value (R *value*)—The R value is an empirical number that indicates the stability or resistance to plastic deformation of a pavement material, and is usually used in the evaluation of base course materials. Marshall size specimens are tested in the stabilometer at room temperature to a vertical pressure of 1.10 MPa (160 psi). The R value is calculated by the following empirical formula

$$R = 100 - \{100/[(2.5/D_2)(P_v/P_h - 1) + 1]\}$$

where

P_v = vertical pressure, 1.10 MPa (160 psi),
P_h = horizontal pressure when P_v is 1.10 MPa (160 psi), and
D_2 = displacement of specimen.

Marshall Stability S_M—The Marshall stability is defined as the maximum load required to produce failure of a standard Marshall specimen in a Marshall test. It is a semi-empirical figure indicating the relative resistance of a material to plastic deformation. The Marshall test was run at room temperature (25°C) instead of the standard 60°C.

Experimental Design

A wide variety of cold-recycled asphalt mixtures were used in this study. The experimental design for this study is presented in Fig. 2. Two different

Pavement Material & Binder / Added Softening Agent	Artificially-Aged Paving Mixture						Old Pavement Material				
	0	.5	1	2	3		0	.5	1	2	3
AE-150	X	X	X	X	X		X	X	X	X	X
Foamed Asphalt	X		X	X	X						
Reclamite	X	X	X								
Mobilsol	X	X	X								
Dutrex 7	X	X	X								

NOTE:

    X Two samples per cell

Tests run: Resilient Modulus test at one day curing and "ultimate" curing condition

R-Value and Marshall tests at "ultimate" curing condition.

Compaction: 60 revolutions at 200 psi (1.38 MPa) with the gyratory machine.

FIG. 2—*Experimental design.*

pavement materials were used: an old pavement material and an artificially aged paving mixture. The artificially aged paving mixtures were used to ensure less variability and better control of the material used. Various levels (up to five levels) of five different softening agents were incorporated in this study. For each mix combination indicated, two specimens were made and tested.

Laboratory Procedure

Specimen Preparation Procedure

The cold recycled asphalt mixtures were prepared in the laboratory. The mixing and curing procedure adopted by the authors in previous studies on cold-recycled mixes was used [3]. This procedure was originally developed by

Gadallah in his study on asphalt emulsion treated mixes [4] and had been used by other researchers [5,6]. The specimen preparation procedure used is described in the Appendix.

Gyratory Compaction

The gyratory machine was used to compact the recycled mixes. The standard procedure as specified in ASTM Test Method for Compaction and Shear Properties of Bituminous Mixtures by Means of the U.S. Corps of Engineers Gyratory Testing Machine (GTM) (D 3387) was generally followed. The initial gyratory angle was set at 1°, and the fixed roller was used. The ram pressure was set at 1.38 MPa (200 psi). The gyrograph recorded during the compaction process was used to obtain the gyratory strain indices. Specimens were compacted with 60 gyratory revolutions.

Testing Procedure

The compacted specimens were left to cure at room temperature, and resilient modulus tests were run on the specimens at one-day curing and at "ultimate" curing condition. "Ultimate" curing condition was obtained by 28-days curing at room temperature followed by a 24-h curing in a forced draft oven at 60°C.

After the resilient moduli had been obtained, the specimens at "ultimate" curing condition were tested in the stabilometer for their R values, and then the same specimens were tested destructively in the Marshall tests.

Results

Gyratory Strain Indices

Artificially Aged Paving Mixtures with AE-150 Added—The gyratory strain indices for the artificially aged paving mixtures with AE-150 added are presented in Fig. 3 as functions of percent AE residue added. It can be noted that the gyratory compactibility indices (GCIs) were insensitive to the changes in percent AE residue added. The gyratory stability index (GSI) remained relatively constant for AE residue added of less than 2% and increased drastically thereafter. The gyratory elastic-plastic index (GEPI) increased as AE residue added increased from 0 to 1%, leveled off as AE residue increased from 1 to 2%, and went up again thereafter.

Results from the resilient modulus, R value, and Marshall tests indicated that the optimum AE residue added was around 1%. Thus it can be noted that GSI and GEPI increased sharply with increase in binder content after the optimum binder content was reached.

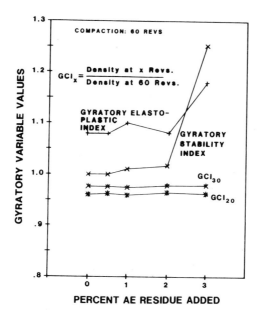

FIG. 3—*Gyratory indices of artificially aged paving mixtures with AE-150 added.*

The gyratory compactibility indices (GCI_{20} and GCI_{30}) were noted to be insensitive to the change in percent AE residue added.

Old Pavement Material with AE-150 Added—The gyratory strain indices for the old pavement material with AE-150 added are presented in Fig. 4. The gyratory compactibility indices (GCIs) were insensitive to the changes in percent AE residue added. The gyratory stability indices (GSI) remained relatively constant for AE residue added of less than 1% and increased greatly thereafter. The gyratory elastic-plastic index (GEPI) increased as AE residue added increased from 0 to 1%, leveled off as AE residue added went from 1 to 2%, and increased again thereafter.

Results from resilient modulus, R value, and Marshall stability tests indicate that optimum AE residue added was around 1%.

Artificially Aged Paving Mixtures with Foamed Asphalt Added—The gyratory strain indices for the recycled mixtures with foamed asphalt added are presented in Fig. 5. It can be noted that the gyratory compactibility indices (GCIs) were insensitive to the changes in percent asphalt added. The stability index increased sharply with increasing binder content when asphalt added was above 2%. The elastic-plastic index increased with increasing percent asphalt added for the entire range.

Results from resilient modulus, R value, and Marshall tests indicated that the optimum asphalt added was around 1%.

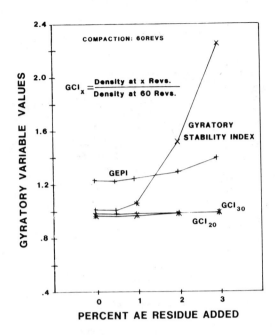

FIG. 4—*Gyratory indices of old pavement material with AE-150 added.*

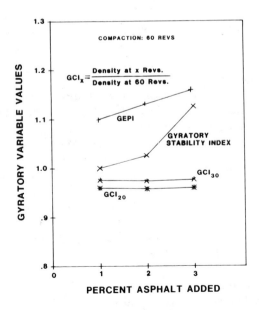

FIG. 5—*Gyratory indices of artificially aged paving mixtures with foamed asphalt added.*

Recycled Mixtures with Rejuvenating Agents Added—Results from the recycled mixtures with rejuvenating agents (Reclamite, Mobilsol, and Dutrex 739) added also indicated that the gyratory compactibility indices (GCIs) were insensitive to the changes in percent agent added. For the small range of percent agent added, no distinct trend can be observed from the gyratory stability index (GSI) and the elasto-plastic index (GEPI).

Comparison Between the Old Pavement Material and the Artificially Aged Paving Mixture—The purpose of using an artificially aged paving mixture in the study was to reduce the material variability so that interrelationships between variables could be observed more easily. In general, the trends observed from the artificially aged paving mixtures were similar to those from the old pavement material. The old pavement material is relatively more sen-

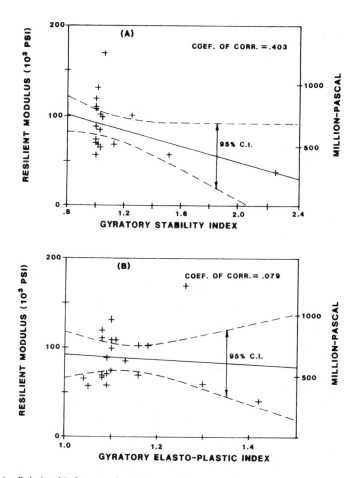

FIG. 6—*Relationship between the resilient modulus at one-day cure and the gyratory indices.*

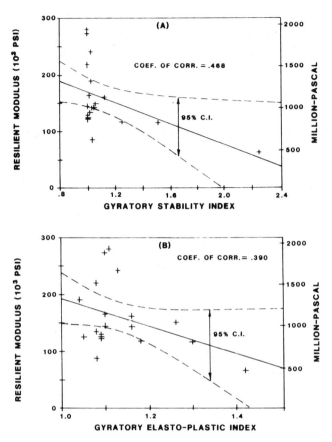

FIG. 7—*Relationship between the resilient modulus at ultimate curing and the gyratory indices.*

sitive to the change in percent added binder, as indicated by the gyratory indices.

Correlation Analysis

Correlation analyses were performed between the gyratory strain indices (GSI and GEPI) and the other response variables (M_R, S_M, and Hveem R values). Results of the analysis indicated that the GSI and GEPI correlated poorly with the resilient modulus and the Marshall stability, but correlated moderately well with the Hveem R value. Figure 6 depicts the plots of resilient modulus at one-day cure versus GSI and GEPI. From these plots, it can be observed that the resilient modulus generally decreased with increasing GSI

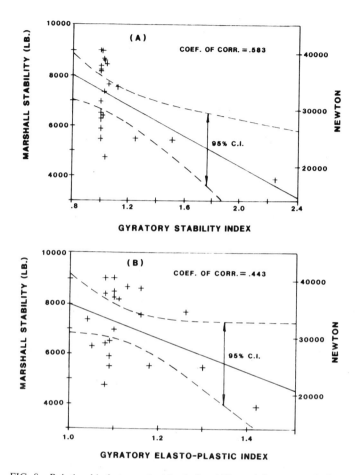

FIG. 8—*Relationship between the Marshall stability and the gyratory indices.*

or GEPI. However, at low values of GSI or GEPI, there were wide ranges of possible corresponding resilient moduli. Figure 7 shows the plots of resilient modulus at ultimate curing versus GSI and GEPI. Similar trends as those in Fig. 6 can be observed. The coefficients of correlations were low, as in the previous case. Figure 8 presents the plots of Marshall stability at ultimate curing versus GSI and GEPI. Similarly, it can be noted that there were wide ranges of possible Marshall stability values at low values of GSI or GEPI. Figure 9 shows the plots of Hveem R value at ultimate curing versus GSI and GEPI. It can be observed that the Hveem R value had a relatively better correlation with GSI and GEPI. The Hveem R value generally decreased with increased GSI or GEPI.

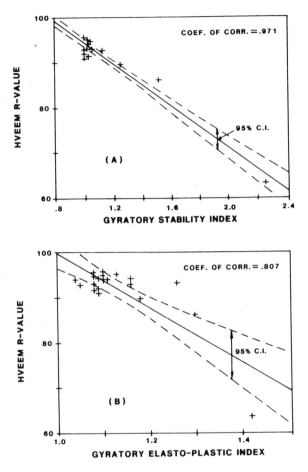

FIG. 9—*Relationship between the Hveem R value and gyratory indices.*

Conclusion

From the results of the analyses presented, it can be concluded that the gyratory stability index and the gyratory elastic-plastic index can be used to detect unstable mixtures when the binder content is too high. When the binder content is near or below the optimum level, the GSI and GEPI are close to 1.0, and they cannot be used to estimate the absolute values of the resilient modulus or the Marshall stability. The R value is relatively insensitive to mix parameters when the mixtures are relatively stable, but becomes very sensitive when the mixtures are relatively unstable. The characteristics of the Hveem R value are similar to those of GSI and GEPI, and thus the Hveem R value has relatively better correlation with GSI and GEPI.

This study is limited to the use of the gyratory testing machine in the fixed roller mode. To explore the full potential of the use of the gyratory testing machine, tests using the oil filled roller mode have to be conducted.

Acknowledgments

The authors wish to express their appreciation to the Indiana Department of Highways and the Federal Highway Administration for their financial support of this study. The asphalt emulsion and rejuvenating agents, Reclamite and Mobilsol, were supplied by K. E. McConnaughay, Inc. The rejuvenating agent Dutrex 739 was supplied by Shell Development Co. The asphalt cements were supplied by American Oil Co. The authors also wish to thank Mr. John L. McRae for the review and the suggestions for revision of this paper.

Appendix

Specimen Preparation Procedure

The cold recycled asphalt mixtures were prepared in the laboratory. The specimen preparation procedure consisted of the following general steps:

1. The proper amount of the pavement material was batched for one specimen.
2. The required amount of water (if any) was added to the material and mixed thoroughly with a mechanical mixer and then with a spoon by hand. The material was then left for 10 to 15 min. This step would be omitted if no added water was required.
3. The proper amount of virgin binder or rejuvenating agent kept at room temperature was added to the material and mixed with a mechanical mixer for $1^1/_2$ min and with a spoon by hand for 30 s.
4. The mix was cured for 1 h in a forced-draft oven at 60°C.
5. The mix was remixed for 30 s with a mechanical mixer and was compacted immediately in the gyratory machine.

The purpose of adding water to the mix was to facilitate the mixing process. When asphalt emulsion, Reclamite, and Mobilsol were used as the added binder, 1% water was added. When foamed asphalt was used, 3% water was added. No water was added when Dutrex was used as the added binder. The optimum amount of water to be added was determined from trial mixes.

References

[1] Tia, M., Wood, L. E., and Hancher, D. E., "The Influence of Certain Factors on the Properties of a Cold-Mix Recycled Asphalt Paving Mixture," *Proceedings, Canadian Technical Asphalt Association*, Vol. 24, Nov. 1979, pp. 78-98.
[2] Schmidt, R. J., "A Practical Method for Measuring the Resilient Modulus of Asphalt-Treated Mixes," Highway Research Board, Record 404, 1972, pp. 22-32.
[3] Tia, M., *A Laboratory Investigation of Cold-Mix Recycled Bituminous Pavements*, Research Report JHRP-78-23, Joint Highway Research Project, Purdue University, West Lafayette, IN, 1978.

[4] Gadallah, A. A., *A Study of the Design Parameters for Asphalt Emulsion Treated Mixtures*, Report JHRP-76-30, Joint Highway Research Project, Purdue University, West Lafayette, IN, 1976.
[5] Iida, A., *The Effects of Added Softening Agents Upon the Behavior of Cold Recycled Asphalt Mixtures*, Report FHWA/IN/JHRP-80/13, Purdue University, West Lafayette, IN, 1980.
[6] Mamlouk, M. S., *Characterization of Cold Mixed Asphalt Emulsion Treated Bases*, Report JRRP-79-19, Purdue University, West Lafayette, IN, 1979.

DISCUSSION

John L. McRae[1] *(written discussion)*—It is unfortunate that the authors did not also have test data using the oil filled roller since this mode of operation of the gyratory testing machine (GTM) is a rational approach to measuring shear strength using plane strain simple shear in which the induced stress is based upon the actual pavement loading and the strain is selected to closely approach prototype conditions. GTM strength parameters and associated moduli, using oil filled roller mode would, of course, be expected to correlate well with stress-strain moduli measured by other means as long as the stress and strain measurements by the other methods related closely to the actual conditions in the pavement structure.

These GTM moduli would not be expected to correlate with the Marshall stability since, despite its extensive use, the Marshall test has the serious handicap of being entirely empirical. Marshall test results are expressed in terms that are not directly applicable in the solution of pavement structural design problems.

The GTM shear and associated moduli, on the other hand, are derived as indicated in Fig. 10 and following.

Gyratory shear S_G is defined as follows

$$2PL = S_G Ah + 2Fa - Vb$$

$$S_G = \frac{2(PL - Fa) + Vb}{Ah}$$

where

P = load on upper roller,
L = lever arm for GTM roller load,

[1] President, Engineering Development Co., Inc., P.O. Box 1109, Vicksburg, MS 39180.

F = force caused by wall friction for $1/2$ of internal mold contact surface,
a = lever arm for load caused by wall friction on $1/2$ of mold,
V = vertical load on test specimen,
b = offset distance for V caused by gyratory shear angle Θ_o,
Θ_o = machine setting for gyratory shear angle,
A = cross-sectional area of test specimen, and
h = vertical height of test specimen.

The gyratory shear modulus G_G is expressed as follows

$$G_G = S_G/\Theta$$

where

G_G = gyratory shear modulus,
S_G = gyratory shear strength, and
Θ = gyratory shear strain.

The gyratory compressive modulus E_G is expressed as follows

$$E_G = 2G_G(1 + \mu)$$

where

E_G = gyratory compressive modulus,
G_G = gyratory shear modulus, and
μ = Poisson's ratio.

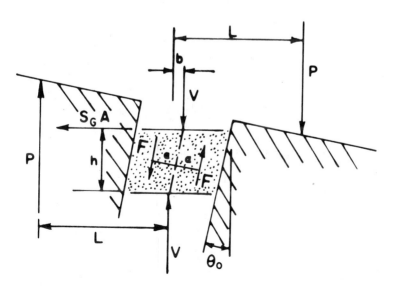

FIG. 10—*GTM force diagram for compaction and shear loading.*

ACTUAL EXAMPLE—STABLE MIX

Crushed Limestone (3/4" Max), 5.5% Bitumen
Tested at $V = 200$ psi Vertical Pressure; $\Theta_o = 1°$ (16 mm); 250°F

Gyratory Shear Modulus $= G_G = \dfrac{S_G}{\Theta_{max}} = \dfrac{78}{0.0197} = 3959$ psi

Gyratory Compression Modulus $= E_G = 2G_G(1 + \mu) = 2 \times 3959 \times 1.5 = 11\,878$ psi

$\epsilon = \dfrac{V}{E_G} = \dfrac{200}{11878} = 0.0168$ in./in.

$\tau_{max} = \dfrac{200}{3.14} = 64$ psi

Gyratory Shear Factor, GSF $= \dfrac{S_G}{\tau_{max}} = \dfrac{78}{64} = 1.22$

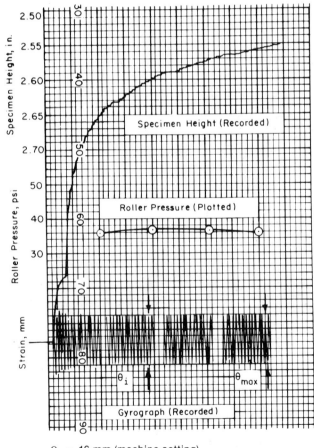

NOTE:
Specimen height curve is becoming asymptotic, indicating approximate compaction equilibrium. Since roller pressure and gyrograph indicate a stable mix, the vertical strain ϵ should be largely recoverable.

NOTE:
No reduction in roller pressure indicates stable condition at this bitumen content.

NOTE:
A gyratory shear factor (GSF) of 1.22 indicates sufficient shear strength to resist shear failure.

NOTE:
No widening of gyrograph indicates stable condition at this bitumen content, also reflected in GSI value of unity.

GSI $= \dfrac{\Theta_{max}}{\Theta_i} = \dfrac{18}{18} = 1.00$

$\Theta_o = 16$ mm (machine setting)
$\Theta_i = 18$ mm (scaled)
$\Theta_{max} = 18$ mm (scaled)

$\Theta_{max} = \dfrac{18}{16} = 1.125° = 1.125 \times 1.75 = 1.969\%$

FIG. 11—*GTM test results.*

DISCUSSION ON A GYRATORY TESTING MACHINE 93

ACTUAL EXAMPLE—UNSTABLE MIX

Crushed Limestone (¾ in. Max), 6.0% Bitumen
Tested at $V = 200$ psi Vertical Pressure; $\Theta_o = 1°$ (16 mm); 250°F

Gyratory Shear Modulus $= G_G = \dfrac{S_G}{\Theta_{max}} = \dfrac{37}{0.0372} = 994$ psi

Gyratory Compression Modulus $= E_G = 2G_G(1 + \mu) = 2 \times 994 \times 1.5 = 2982$ psi

$$\epsilon = \frac{V}{E_G} = \frac{200}{2982} = 0.0670 \text{ in./in.}$$

$$\tau_{max} = \frac{200}{3.14} = 64 \text{ psi}$$

Gyratory Shear Factor, GSF $= \dfrac{S_G}{\tau_{max}} = \dfrac{37}{64} = 0.58$

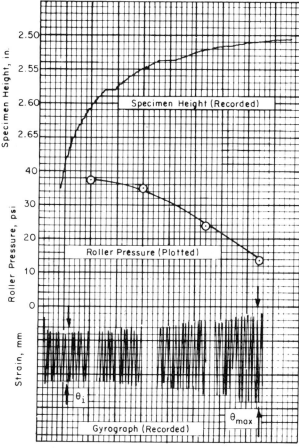

NOTE:
Specimen height curve is becoming asymptotic, indicating approximate compaction equilibrium. However, progressive rut development would be expected because roller pressure is dropping and gyrograph is spreading. The high value of ϵ is an indication that the accumulative rut depth would be large.

NOTE:
Progressive reduction in GTM roller pressure warns of instability at this bitumen content.

NOTE:
A gyratory shear factor (GSF) of 0.58 indicates theoretical shear stress seriously in excess of shear strength and therefore shear failure would be anticipated.

NOTE:
Progressive widening of gyrograph warns of instability at this bitumen content. This is reflected in GSI value greater than unity.

$\Theta_o = 16$ mm (machine setting)
$\Theta_i = 19$ mm (scaled)
$\Theta_{max} = 34$ mm (scaled)

$$\text{GSI} = \frac{\Theta_{max}}{\Theta_i} = \frac{34}{19} = 1.789$$

FIG. 12—*GTM test results.*

$$\Theta_{max} = \frac{34}{16} = 2.125° = 2.125 \times 1.75 = 3.719\%$$

The gyratory compressive strain ϵ is expressed as follows

$$\epsilon = V/E_G$$

where

V = vertical load and
E_G = gyratory compressive modulus.

When the induced stress and strain simulate that in the pavement structure, then the "pavement density" as well as the stress-strain parameters relate rationally to the structural design problem.

The GTM combination of kneading compaction with a plane strain simple shear test is uniquely related to pavement design requirements in that the design density requirement as well as the strength requirement are both rationally adjusted to changes in the traffic loading.

Figures 11 and 12 show actual test results for a stable pavement mix and an unstable pavement mix. The GTM shear test can be conducted at lower temperatures than the 121°C (250°F) used here, and the test will then, of course, reflect higher shear strength and larger moduli because of the cohesive property of the bitumen. The writer considers the shear test at the compaction temperature to be an expedient measure of the effective shear strength that is due primarily to the internal friction and interlock of the aggregate.

It cannot be too strongly emphasized that the optimum mix design is a function of the degree of compaction required for the anticipated traffic loading. For this reason it is necessary to simulate the design loading and compact to an equilibrium density condition in the GTM when measuring the shear strength and related moduli.

Mang Tia and Leonard E. Wood (authors' closure)—The authors wish to express their sincere appreciation for Mr. McRae's valuable discussion on the paper. The results of this study indicate that the use of the gyratory testing machine in the fixed roller mode will be limited to the determination of optimum asphalt contents. The GTM shear test using the oil filled roller is a rational way to measure the shear strength of an asphalt mixture at specified temperature and density. The GTM has been used widely as a compaction machine. Its great potential as a rational testing machine needs to be fully explored.

Index

A

AASHTO Standard T 242, 38
Anionic asphalt emulsion, 77
Asphalt cement, AC 2.5, 77
Asphalt polymer, 66, 68
ASTM standards
  D 1856, 77
  D 3387, 82
  E 274, 38
  E 503, 37
  E 556, 39
  E 670, 36

B

Balmer, Glenn G., 33-50

C

Chong, George J., 3-17
Cold-mix recycling, 75
Condition survey, highway, 4
Cost effectiveness of maintenance treatment, 8
Crack sealing
  Guidelines, 14
  Methods and materials, 14, 15
  Rationale, 11, 12
  Research and improvement, 15

D

Deen, Robert C., 18-32
Deflections, dynamic
  Deflection bowl, 21
  Errors, 23
  Kentucky Road Rater, 20
  Reference conditions, 21
  Sampling and statistics, 28

E

Expected life of maintenance alternatives, 7, 8

F

Friction characteristics, remedial procedures
  Grooving, 47, 48
  Milling, 46, 47
  Open-graded friction course, 48, 49
  Resurfacing, 48
  Sprinkle treatment, 48
Friction, pavement measurement methods
  Articulation angle, 41, 42
  At intersections, 43
  Diagonal braked vehicle, 37, 38
  Dynamic vertical test-wheel load, 43, 44
  Mu-meter, 36, 37
  Saab friction tester, 45, 46
  Side force measurement transducer, 39
  Single wheel tester, 44, 45

G

Gyratory indices
  Compactibility, 80

Elasto-plastic, 79
Stability, 79
Gyratory testing maching, 75, 76

O

Overlay design, 29

H

Hoffman, Gary L., 51-64

P

Pavement condition rating, 10
Phang, William A., 3-17
Preventive maintenance, 10

K

Kallas, Bernard F., Editor, 1, 2
Knight, Norman E., 51-63

R

Recycling agents, 77
Rowlett, David R., 64-73

M

Maintenance guidelines, pavement, 4
Maintenance performance standards, 7
Marshall stability, 80
Membranes
  Asphaltic mastic and nonwoven polypropylene fabric, 54
  Fiberglass mesh polymer modified asphalt and nonwoven polypropylene fabric, 54
  Fiberglass woven fabric and polymerized asphalt, 54
  Polypropylene woven mesh and self-adhesive rubberized asphalt, 54
  Rubberized asphalt impregnated nonwoven polypropylene fabric, 54
  Rubberized asphalt with polypropylene mesh, 52
Moduli, asphalt concrete, 24

S

Sharpe, Gary W., 18-32
Southgate, Herbert F., 18-32
Stabilometer resistance value, 80
Subgrade strength, 25

T

Tia, Mang, 74-94
Tonda, Richard D.,

U-Z

Uffner, William E., 64-73
Wood, Leonard E., 74-94
Woven glass fiber fabric, 66
Zimmer, Richard A., 33-50